Synthesis Lectures on Electrical Engineering

This series of short books covers a broad spectrum of titles of interest in electrical engineering that may not specifically fit within another series. Books will focus on fundamentals, methods, and advances of interest to electrical and electronic engineers.

Gareth Monkman

Digital Electronics

 Springer

Gareth Monkman
Department of Electrical Engineering
OTH-Regensburg
Regensburg, Germany

ISSN 1559-811X   ISSN 1559-8128 (electronic)
Synthesis Lectures on Electrical Engineering
ISBN 978-3-031-69725-8   ISBN 978-3-031-69726-5 (eBook)
https://doi.org/10.1007/978-3-031-69726-5

© The Editor(s) (if applicable) and The Author(s), under exclusive license to Springer Nature Switzerland AG 2026

This work is subject to copyright. All rights are solely and exclusively licensed by the Publisher, whether the whole or part of the material is concerned, specifically the rights of translation, reprinting, reuse of illustrations, recitation, broadcasting, reproduction on microfilms or in any other physical way, and transmission or information storage and retrieval, electronic adaptation, computer software, or by similar or dissimilar methodology now known or hereafter developed.
The use of general descriptive names, registered names, trademarks, service marks, etc. in this publication does not imply, even in the absence of a specific statement, that such names are exempt from the relevant protective laws and regulations and therefore free for general use.
The publisher, the authors and the editors are safe to assume that the advice and information in this book are believed to be true and accurate at the date of publication. Neither the publisher nor the authors or the editors give a warranty, expressed or implied, with respect to the material contained herein or for any errors or omissions that may have been made. The publisher remains neutral with regard to jurisdictional claims in published maps and institutional affiliations.

This Springer imprint is published by the registered company Springer Nature Switzerland AG
The registered company address is: Gewerbestrasse 11, 6330 Cham, Switzerland

If disposing of this product, please recycle the paper.

Acknowledgments

Many thanks to all those who have helped make this book possible including colleagues and students for helpful feedback. A special thanks to the sources for schematics and photos, particularly the Radio Museum in Cham for much inspiration.

Contents

1 Introduction to Digital Electronics 1
   1.1   Introduction 1
   1.2   Devices 1
          1.2.1   Diodes 2
          1.2.2   Transistors 5
   References 9

2 Combinational Logic 11
   2.1   Time Delay Effects 14
   2.2   Logic Symbols and Representations 15
   2.3   Rules of Boolean Algebra 15
          2.3.1   Extension to De Morgan's Rule 16
   2.4   Truth Tables 17
   2.5   Karnaugh Maps 17
   2.6   Sum of Products and Product of Sums 20
   2.7   Transmission Gates and Tri-State 22
   2.8   Multiplexers 23
   2.9   Canonical Form 24
   2.10 Programmable Logic Devices 27
   2.11 PROM (Programmable Read-Only Memory) 27
   2.12 PAL (Programmable Array Logic) 28
   2.13 PLA (Programmable Logic Array) 28
   2.14 GAL (Generic Array Logic) 29
   2.15 FPGA (Field Programmable Gate Arrays) 29
   2.16 CPLD (Complex Programmable Logic Device) 30
   2.17 Example VHDL Code for OR and AND Functions 30
   2.18 Binary Hardware Addition and Subtraction 32
   2.19 Subtractor 33
   2.20 Comparators 35

	2.21	Encoders and Decoders	36
	2.22	Arithmetic and Logic Units (ALU)	38
	2.23	Digital Simulators	40
		Appendix—Logic Families	41
		Exercises 2—Combinational Logic	43
		References	52
3	**Sequential Logic**	55
	3.1	SR Flip Flop	55
	3.2	Latches	57
		3.2.1 The D-type Flip-Flop	58
		3.2.2 The T Flip-Flop (Toggle Flip-Flop)	58
	3.3	The JK Flip Flop	60
		3.3.1 Dual JK Flip-Flop 74LS73	63
		3.3.2 The Master–Slave JK Flip-Flop	64
	3.4	Digital Counters	65
		3.4.1 Asynchronous Counters	65
		3.4.2 Synchronous Counters	68
		3.4.3 Shift Registers	71
	3.5	State Diagrams	75
		Exercises 3—Sequential Logic	78
		References	82
4	**Binary Mathematics**	83
		Exercises 4—Binary Mathematics	92
		References	97
5	**The Basic Computer**	99
	5.1	Hardware	99
	5.2	Arithmetic and Logic Unit (ALU)	101
		Exercises 5–Micro–Computer Systems	106
		References	106
6	**Software**	107
	6.1	The Asm Command in C	108
	6.2	Basic Simple Assembler	110
	6.3	Compilers	116
	6.4	Example of Lexical Analysis, Tokens, Non-Tokens	116
	6.5	Examples of Tokens Created	117
	6.6	Examples of Nontokens	117
	6.7	Interpreters	118
		Exercises 6–Assembler	118
		References	119

7 Applications ... 121
- 7.1 Gray Code ... 121
  - 7.1.1 Binary–Gray and Gray–Binary Conversions ... 122
  - 7.1.2 Incremental Counters ... 124
- 7.2 Parallel Bus Systems ... 125
  - 7.2.1 7-Segment Display ... 128
- 7.3 Error Detection ... 128
  - 7.3.1 Parity Checking ... 129
  - 7.3.2 Differential Manchester Encoding ... 131
  - 7.3.3 Programmable Logic Controllers (PLC) ... 132
- Exercises 7–Applications ... 134
- References ... 135

8 A Short History of Computers ... 137
- 8.1 A Short History of Calculating Machines ... 138
- 8.2 Memory ... 142
- 8.3 Memory Cells ... 143
- 8.4 Non-Volatile Memory ... 146
- 8.5 Memory Arrays ... 148
- 8.6 A Compact History of the Microprocessor ... 153
- References ... 154

Introduction to Digital Electronics

1.1 Introduction

Although the details of analogue electronics are often first taught, or at least parallel to digital electronics, the introductory parts should suffice for a basic understanding of the devices used in digital circuitry. As a prerequisite, only basic mathematics, engineering knowledge and fundamental electrical theory (such as Ohm's and Kirchhoff's laws) are required.

The text introduces fundamental digital components and devices, often with respect to historical events which led to their conception, before combining them to design complete digital systems. The historical aspect is important in that many older devices were purely mechanical and demonstratively simple to comprehend. This helps students to understand basic concepts before embarking on analysis and application of the more covert modern semiconductor devices.

References are provided, not only for information sources directly relevant to the text, but also to more advanced texts. This is intended to serve as a bridge between this basic introductory course for first semester (first year) undergraduates and later courses. General programming languages are mentioned and occasional references to C are made, but without great detail as computer programming is usually part of a parallel information technology course.

1.2 Devices

Analog electrical signals can have a theoretically infinite number of levels as opposed to digital signals which can have only two levels thus representing the two binary states of "on" and "off". Although many analogue devices can be used in digital form by driving them to their extremes, most digital components are designed to handle only digital signals very quickly and cannot be used as analogue devices.

1.2.1 Diodes

Diodes are simply electronic devices which allow current to flow in only one direction. The very first diodes, as shown in Fig. 1.1, exploited the non-linear characteristics of quartz and galena crystals as used in very early radios (crystal sets) for rectification of modulated radio frequency signals.

Finding the right point on the crystal with a piece of wire known as the cat's whisker wasn't easy but it led to a condition where current could pass in only one direction. How this principle is used in modern semiconductor diodes will be revealed later.

Have you ever wondered why iron rusts quicker when it is heated? Metals have free electrons on their surface which are emitted when heated. Most simply fall back but in the meantime oxygen atoms have a chance to combine with the iron – oxidation. When a wire is heated by passing a current through it, the same happens. However, if that wire is in a vacuum then it cannot oxidize and the emitted electrons can be attracted to a positively charged plate (an anode). Consequently, current can flow from the wire (cathode) to the anode but not the other way around. This is how the first vacuum tube diodes (Thermionic Valve) worked as depicted in Fig. 1.2a.

Fig. 1.1 Crystal detector (Courtesy: Rundfunkmuseum, Cham, Germany)

1.2 Devices

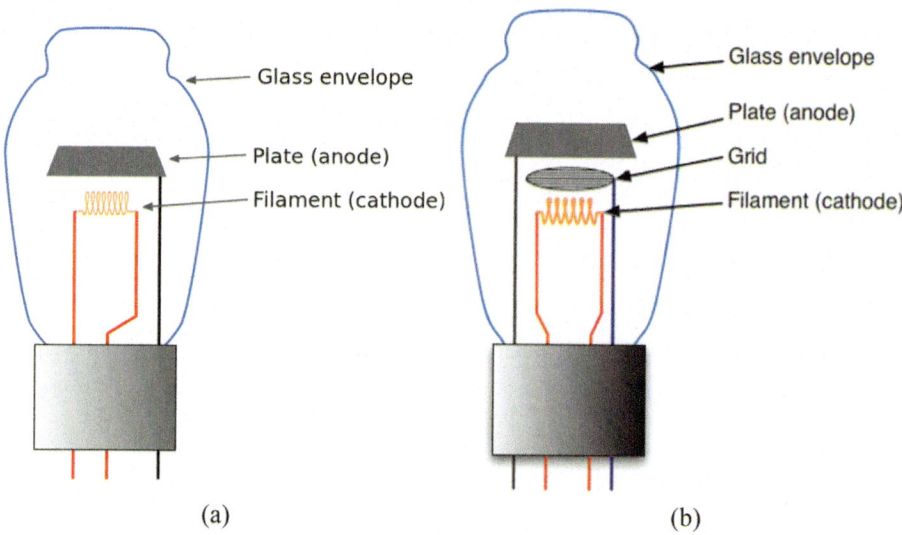

Fig. 1.2 Thermionic valve principle. **a** diode; **b** triode (Courtesy: creativecommons.org)

Adding a third electrode in the form of a grid between the cathode and anode, as shown in Fig. 1.2b, makes it possible to control the electron flow. This is known as a triode and works in the same manner as a field effect transistor. This will be considered again in semiconductor form later.

The modern version of the old thermionic diode valve is the semiconductor p–n junction diode. When forward biased, as shown in Fig. 1.3, electric current is allowed to flow in only one direction while current in the opposite or reverse direction (when reverse biased, as shown in Fig. 1.4) is blocked. This is similar to the old quartz crystal, but the built-in p–n junction obviates the need to fish around with a cat's whisker.

In n-type semiconductors, free electrons are the majority charge carriers whereas in p-type semiconductors, holes are the majority charge carriers. When the n-type semiconductor is joined together with the p-type semiconductor, a p–n junction is formed.

The p–n junction diode is made from semiconductor materials such as silicon, germanium or gallium arsenide. The introduction of impurities into the lattice structure of semiconductor crystals, a process known as doping, changes the electrical conductivity. P-type Silicon is usually doped with a tri-valent material such as Aluminum, Boron, Gallium or Indium and for n-type doping penta-valent materials such as Arsenic, Antimony or Phosphorus and employed [1].

Most modern p–n junction diodes are made from silicon semiconductors which works at higher temperatures when compared with the p–n junction diodes made from germanium. The advantage of Germanium is the lower band gap energy making the room temperature forward biased voltage drop lower (ca. 0.16 V) than that for silicon (ca.

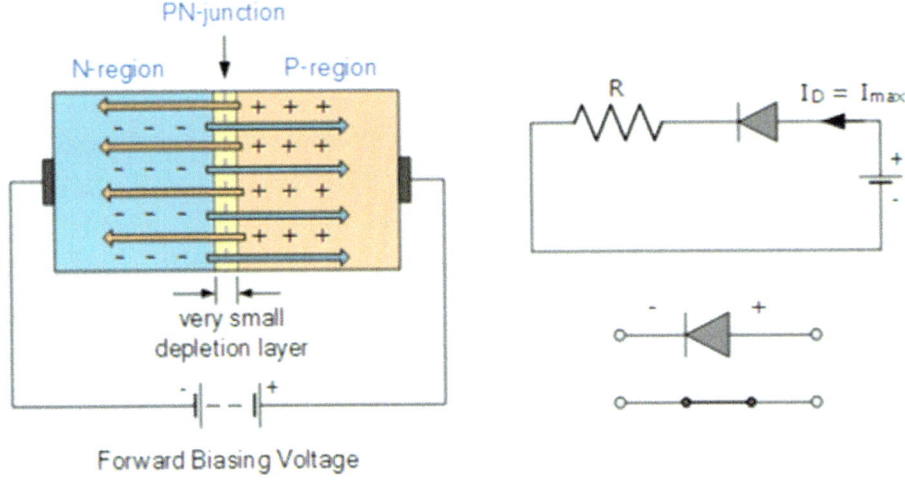

Fig. 1.3 Semiconductor diode p–n junction forward biased (Courtesy: Studytronics)

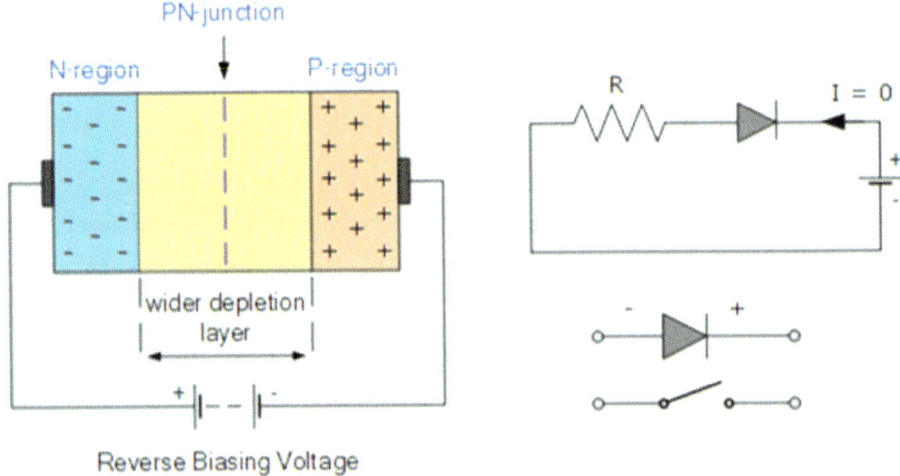

Fig. 1.4 Semiconductor diode p–n junction reverse biased (Courtesy: Studytronics)

0.7 V). This is advantageous for radio receivers where signals are small but for a digital circuit supplied with 5 V, such potential drops are not so important.

1.2.2 Transistors

There are basically two types of transistor. The field effect transistor is a voltage controlled device which works almost exactly as a thermionic valve but without the need for a heater and usually at much lower voltages. In fact, field effect transistors can be used to replace defective valves in vintage radio receivers! The second version, the bipolar transistor, is a current driven device. Both will be explained as we proceed.

1.2.2.1 Field Effect Transistors
In the same way as the thermionic valves of the past can be controlled by means of a voltage on the grid, field effect transistors are controlled by voltage changes on the gate. Although they are essentially analogue devices, they can be used as simple switches in digital electronics by driving them fully on or fully off.

1.2.2.2 The MOSFET
The MOSFET (Metal–Oxide–Semiconductor Field-Effect Transistor) is a particularly low-loss field-effect transistor with a MOS structure. MOSFETs are typically three-terminal devices with drain (D), source (S) and gate (G) terminals, analogous to the anode (A), cathode (C) and grid (G) of the thermionic valve. In addition, there are four-terminal power FET devices with an extra gate [2] or sense terminal [3]. Current flow between drain (D) and source (S) is controlled by a voltage applied to the gate (G). MOSFETS can be P-channel or N-channel and be enhancement or depletion types (Fig. 1.5).

The enhancement type MOSFET requires a Gate-Source voltage, (V_{GS}) to switch the device "ON" which makes the enhancement mode MOSFET equivalent to a "Normally Open" switch. The depletion type MOSFET requires a Gate-Source voltage, (V_{GS}) to switch the device "OFF" which makes the depletion mode MOSFET equivalent to a "Normally Closed" switch.

1.2.2.3 The Bipolar Transistor
Contrary to field effect transistors, Bipolar Transistors are driven by current not by voltage. This often makes them more difficult to understand. There are PNP and NPN versions. With PNP transistors, the emitter is connected to the positive supply terminal and the collector to the negative side. A very small amount of current flowing between the collector and emitter is siphoned off through the base. Controlling this current (which is usually a few micro-amps) controls the much larger collector-emitter current (which can be in the mA to Ampere range). With NPN transistors the emitter is negative and the collector positive. A small current flowing into the base allows the collector-emitter current to be controlled in the same manner (Fig. 1.6).

The relationship between base current and collector current is effectively the maximum gain of the device and is usually denoted as β or HFE (or hfe where small signal currents are concerned).

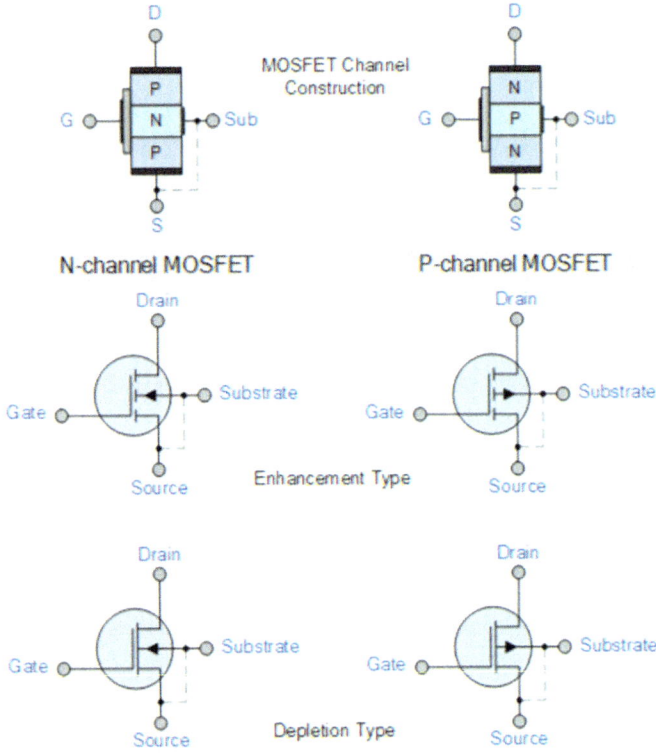

Fig. 1.5 MOSFET principle (Courtesy: Electronics-tutorials)

$$\alpha = \frac{I_C}{I_E} \text{ and } \beta = \frac{I_C}{I_B} \text{ therefore } I_C = \alpha I_E = \beta I_B$$

$$\text{so } \alpha = \frac{\beta}{1+\beta} \text{ and } \beta = \frac{\alpha}{1-\alpha}$$

Both field effect and bipolar transistors are basically analogue devices which are driven to being fully on or fully off when used in digital circuits. When a bipolar transistor is fully switched on, it acts like a simple switch, as in the previous example using FETs (Fig. 1.7).

Because of the relatively large base-emitter voltage (typically 0.7 Volts at room temperature) bipolar transistors tend to be slow when switched. However, the Schottky diode (sometimes called a "hot carrier diode") experiences less than half this forward voltage drop making switch-off faster [4].

1.2 Devices

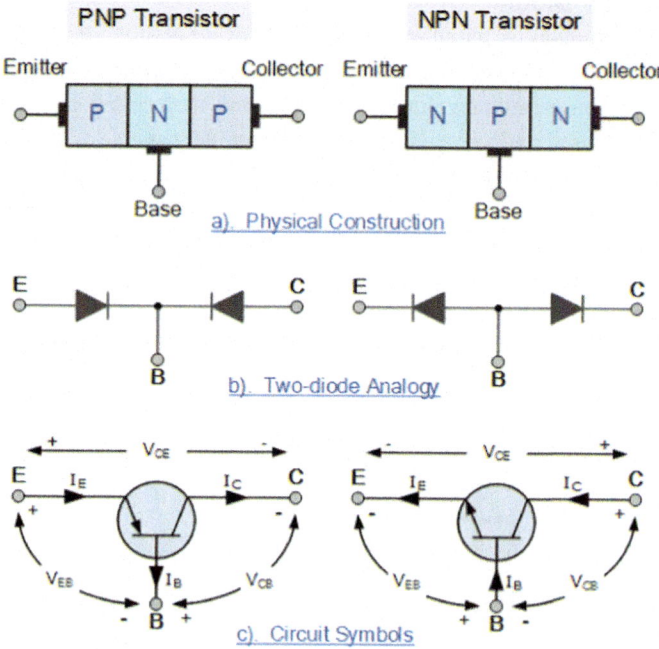

Fig. 1.6 Bipolar transistor principle (Courtesy: Electronics-tutorials) {permission granted 231,204}

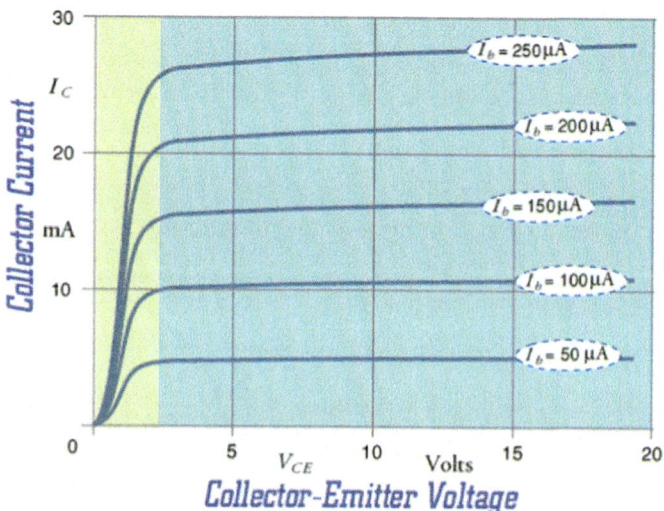

Fig. 1.7 Bipolar transistor characteristics (Courtesy: commons.wikimedia.org)

1.2.2.4 Schottky Diode

Schottky TTL technology is based on the insertion of a Schottky diode that bridges the base and collector of a bipolar transistor, which is then referred to as a "Schottky transistor". When the base-emitter voltage is positive and the transistor is conducting, the diode is forward-biased and prevents the collector-emitter voltage from dropping below around 0.3–0.4 V. The transistor does not reach full saturation, resulting in a faster switching time than with standard TTL devices (Fig. 1.9). Transistors with integrated Schottky diodes have their own symbol as shown in Fig. 1.8b—though it is not always used today!

Because the Schottky diode prevents the bipolar transistor from going completely into saturation, the need for a long recovery time at switch-off is eliminated. When a transistor is fully on or fully off, the power losses are minimized. Most power dissipation occur

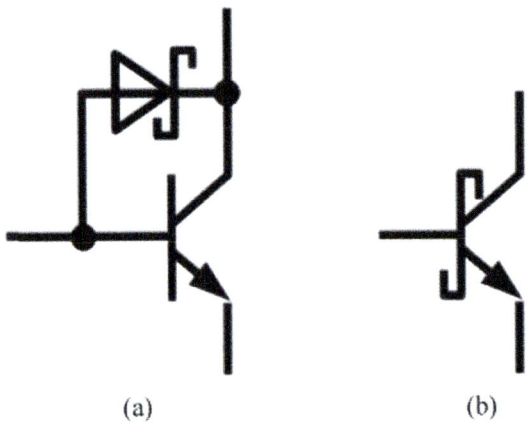

Fig. 1.8 a Switching acceleration by means of Schottky diode. b Schotty diode symbol

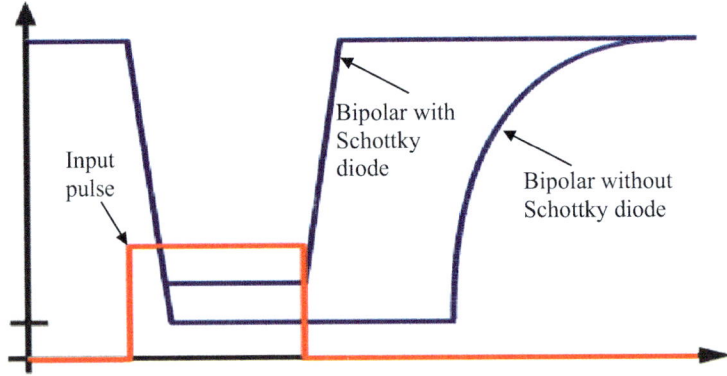

Fig. 1.9 Comparison of bipolar transistor with and without Schottky diode

during the time it takes to switch (rise and fall times). Consequently, the faster a transistor switches, the lower the power losses.

For greater detail with respect to such devices, the reader should consult texts dealing with both analog and digital aspects of devices [5].

References

1. Solymar L., Walsh D. (1993) Lectures on the electrical properties of materials. Oxford University Press. https://inis.iaea.org/search/search.aspx?orig_q=RN:26065836
2. Leistiko O. and D. Hilbiber, "The FET tetrode: A high-performance, small-signal amplifier," *1964 International Electron Devices Meeting*, Washington, DC, USA, 1964, pp. 60-60, https://doi.org/10.1109/IEDM.1964.187473.
3. Onsemi—*Current Sensing Power MOSFETs*—Semiconductor Components Industries, LLC.—Publication Order Number: AND8093/D, March, 2017—Rev. 6. www.onsemi.com
4. Biard, James R., "*Unitary Semiconductor High Speed Switching Device Utilizing a Barrier Diode*", US Patent US 3463975, filed December 31, 1964, issued August 26, 1969.
5. Storey. N.—*Electronics: A Systems Approach* – Addison-Wesley, Harlow UK, 1998.

Internet Sites

6. https://chamer-rundfunkmuseum.de/
7. http://www.historywebsite.co.uk/Museum/Engineering/Electronics/history/earlytxrx.htm
8. http://www.physics-and-radio-electronics.com/electronic-devices-and-circuits/semiconductor-diodes/pnjunctionsemiconductordiode.html
9. http://www.electronics-tutorials.ws/
10. http://electronicsinourhands.blogspot.de/2014/
11. https://studytronics.weebly.com/

Combinational Logic 2

Combinational logic is the concatenation of digital electronic devices, usually without feedback loops.

Since their inception during the late 1950s, the original RTL (resistor-transistor logic), DTL (diode-transistor logic) and TTL (transistor-transistor logic) have been improved or replaced by faster devices having lower energy consumption. In fact, modern digital systems can switch well into the GHz range. Many devices contain several gates as shown in the example of a 7400 quad NAND-gate in Fig. 2.1.

Pin 14 is used for the 5 Volt supply and pin 7 for ground (0 Volts). This is usual for most 14-pin TTL devices of the 74-series.

The simplest digital device comprises an OR gate made from two inputs connected together through diodes. Adding a transistor causes the state to be inverted giving a DTL NAND gate.

The circuit shown in Fig. 2.2 depicts a DTL NAND gate with two inputs and one output. From the truth table for the NAND function, depicted in Table 2.1, it can be seen how the circuit works. If any input is taken to ground (logic 0) then the transistor VT1 is switched off and the Output will be taken high (logic 1) by the resistor R2. Should 5 Volts (Vcc) appear on both inputs then the transistor will be switched on by resistor R1 thus pulling the output down to ground (logic 0). It is interesting to note that if nothing is connected to the inputs then the resistor R1 will cause the transistor to remain switched on.

In the TTL version of the NAND gate shown in Fig. 2.3, a special form of transistor (not normally available in discrete device form) with two emitters is used. Because the base of Q1 is high, should any input be logic 0 then Q1 switches on, pulling the base of

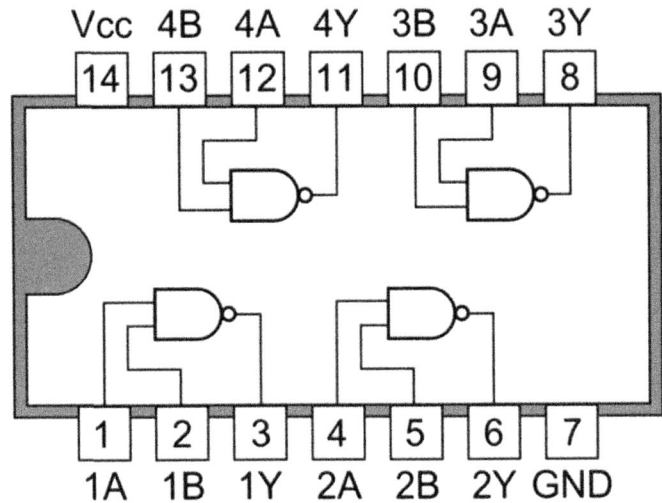

Fig. 2.1 7400 quad NAND gate integrated circuit (Courtesy: creativecommons.org)

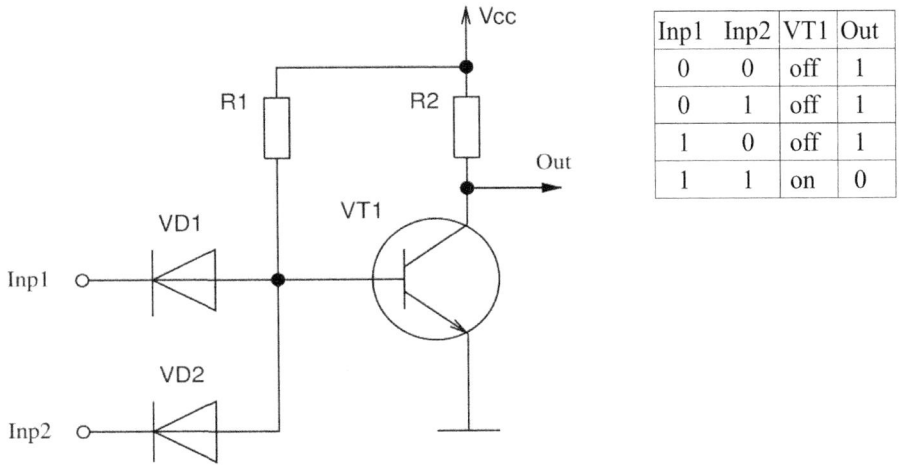

Fig. 2.2 DTL NAND gate (Courtesy: creativecommons.org)

Q2 to ground and thus switching it off. With Q2 off, R2 causes Q3 to switch on and at the same time R3 turns Q4 off. This results in a logic 1 at the output.

Conversely, if both inputs are high, Q1 to Q4 reverse their conditions and logic 0 appears at the output. The diode D1 simply prevents current flowing, from whatever is connected to the output, into transistor Q4.

2 Combinational Logic

Table 2.1 Gate symbols and representations

Function	ANSI Symbol	IEC Symbol	Boolean	Truth table			Karnaugh map		
NOT			$F = \overline{A}$	A		F			
				0		1			
				1		0			
AND			$F = A \cdot B$	A	B	F	$_B^A$	0	1
				0	0	0	0	0	0
				0	1	0	1	0	1
				1	0	0			
				1	1	1			
OR			$F = A + B$	A	B	F	$_B^A$	0	1
				0	0	0	0	0	1
				0	1	1	1	1	1
				1	0	1			
				1	1	1			
XOR			$F = A \oplus B$	A	B	F	$_B^A$	0	1
				0	0	0	0	0	1
				0	1	1	1	1	0
				1	0	1			
				1	1	0			
NAND			$F = \overline{A \cdot B}$	A	B	F	$_B^A$	0	1
				0	0	1	0	1	1
				0	1	1	1	1	0
				1	0	1			
				1	1	0			
NOR			$F = \overline{A + B}$	A	B	F	$_B^A$	0	1
				0	0	1	0	1	0
				0	1	0	1	0	0
				1	0	0			
				1	1	0			
XNOR			$F = \overline{A \oplus B}$	A	B	F	$_B^A$	0	1
				0	0	1	0	1	0
				0	1	0	1	0	1
				1	0	0			
				1	1	1			

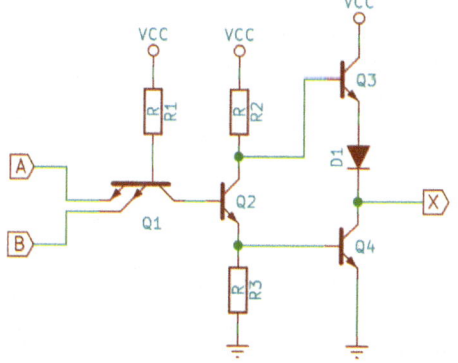

A	B	Q1	Q2	Q3	Q4	X
0	0	on	off	on	off	1
0	1	on	off	on	off	1
1	0	on	off	on	off	1
1	1	off	on	off	on	0

Fig. 2.3 TTL NAND gate (Courtesy: Robert Baruch)

For input levels, logic 0 in standard 5 Volt TTL is defined as all voltages between 0 and 0.8 Volts and logic 1 between 2 and 5 Volts. For output levels, logic 0 is defined as all voltages between 0 and 0.4 Volts and logic 1 between 2.7 and 5 Volts. All other potentials between are not defined! For modern 3.3 Volt TTL the undefined regions are 0.8 to 2 Volts for inputs and 0.5 to 2.4 for outputs [1].

The number of gate outputs which can be connected to an individual gate input in known as "fan-in". The amount of gate inputs which can be driven from a single gate output in known as "fan-out".

Conventionally, should a DTL or TTL gate input be left unconnected it will normally act as logic high. However, to avoid uncertainties all unused logic inputs should be wired high (or low). With CMOS devices unconnected inputs MUST be connected high (or low) as their state is otherwise undefined and their high input impedances makes them very sensitive to noise!

In addition to working voltages and current, temporal parameters must be considered. Timing is a major consideration when designing any digital system.

2.1 Time Delay Effects

Truth tables provide a somewhat idealized representation. Real gates also have time delays which must also be taken into account. As already explained, digital devices are fast but not infinitely so. As depicted below, the switching takes time. The hysteresis between on and off results in time delays which can accumulate in large circuits. All changes in the state of a gate are initiated by the rising (or falling) edge of the input signal. However, such changes do not take place instantly as it takes time for the current to flow—propagation delay!

Figure 2.4 shows an example for a digital inverter where the falling edge of the output follows the rising edge of the input with a delay of t_{PHL}. Similarly, when the input changes to logic 0, the output changes to logic 1 after a delay t_{PLH}. Note that these time delays can be unequal, as depicted in Fig. 2.4.

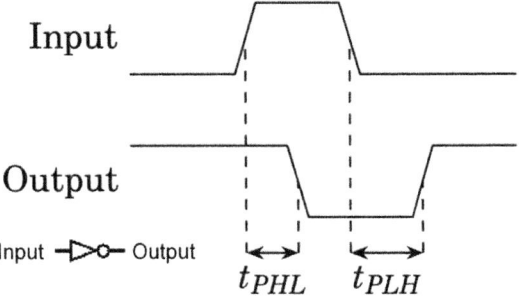

Fig. 2.4 Timing diagrams showing delay between input and output (Courtesy: creativecommons.org)

2.3 Rules of Boolean Algebra

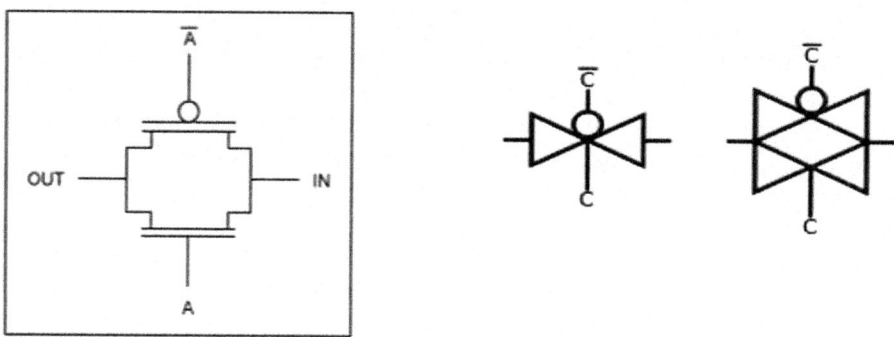

Fig. 2.5 Transmission gate schematic and commonly used symbols

Other common timing problems are *hazards*. *Race* hazards occur when there are two paths through a circuit with unequal propagation delay. *Static* hazards come in the form of "glitches" commonly associated with double pulses often resulting from key bounce [2]. The effects of propagation delays will be considered in greater detail in sequential logic!

2.2 Logic Symbols and Representations

Both the original (and more widespread) US Mil-Spec (ANSI) symbols and the modern European symbols, shown in Table 2.1, are deliberately used in this book as the student should become used to both systems. The function of each type of gate may be described by accompanying truth tables and Karnaugh maps.

Sometimes the triangle denoting inversion in the IEC symbol is replaced with a circle as in the ANSI symbol.

2.3 Rules of Boolean Algebra

All such gates obey the rules of Boolean algebra. The following are the 25 rules [2] of Boolean algebra [3]. Many of them are fairly obvious, the more complicated rules will be proven using various techniques throughout this book.

1. $A + \overline{A} = 1$
2. $A + A = A$
3. $A \cdot A = A$
4. $A \cdot \overline{A} = 0$

5. $A + 0 = A$
6. $A + 1 = 1$
7. $A \cdot 1 = A$
8. $A \cdot 0 = 0$
9. $A \cdot B = B \cdot A$ Commutative law
10. $A + B = B + A$ Commutative law
11. $A \cdot (B + C) = A \cdot B + A \cdot C$ Distributive law
12. $A + B \cdot C = (A + B) \cdot (A + C)$ Distributive law
13. $A + B + C = (A + B) + C = A + (B + C)$ Associative law
14. $A \cdot B \cdot C = A \cdot (B \cdot C) = (A \cdot B) \cdot C$ Associative law
15. $A \cdot (B + \overline{B}) = A$
16. $A + A \cdot B = A$ Absorption law
17. $A \cdot (A + B) = A$ Absorption law
18. $A + \overline{A} \cdot B = A + B$
19. $B \cdot (A + \overline{B}) = A \cdot B$
20. $(A + B) \cdot (B + C) \cdot (C + \overline{A}) = (A + B) \cdot (C + \overline{A})$
21. $\overline{A + B} = \overline{A} \cdot \overline{B}$ De Morgan's rule
22. $\overline{A \cdot B} = \overline{A} + \overline{B}$ De Morgan's rule
23. $A \cdot B + B \cdot C + \overline{A} \cdot C = A \cdot B + \overline{A} \cdot C$
24. $\overline{A} + A \cdot B = \overline{A} + B$
25. $A \cdot B + A \cdot C + \overline{B} \cdot C = A \cdot B + \overline{B} \cdot C$

[2]

2.3.1 Extension to De Morgan's Rule

$$\overline{A + B + C} = \overline{A} \cdot \overline{B} \cdot \overline{C} \text{ and } \overline{A \cdot B \cdot C} = \overline{A} + \overline{B} + \overline{C}$$

Boolean algebra may be used to minimise the number of gates in a circuit design. Other design techniques include Karnaugh-Veitch diagrams (also known as Karnaugh maps).

2.4 Truth Tables

Truth tables are a very simple way of illustrating the function of digital circuits.

A	B	F
0	0	0
0	1	1
1	0	1
1	1	1

The inputs A and B lead to the output F via a pre-defined function, in the above case logical OR. Another method is the Karnaugh map.

2.5 Karnaugh Maps

Karnaugh Maps are variations of the Venn diagram used to clearly represent and simplify a Boolean function into minimal a logical expression.

$_B{}^A$	0	1
0	0	1
1	1	1

In the table below both truth tables and Karnaugh maps can be seen for the 5 most common logic functions (Table 2.2).

Karnaugh Maps can be extended to 3 or 4 variables. After this it becomes a bit difficult.

Note the topology is not exactly 1:1 with that of the truth table. For a 3 variable Karnaugh map A and B are in the order: 00 01 11 10 and *NOT* as with truth tables: 00 01 10 11.

$_C{}^{AB}$	00	01	11	10
0				
1				

For example, the function: $F = \overline{A} \cdot B + \overline{A} \cdot \overline{B} \cdot \overline{C} + A \cdot B \cdot \overline{C} + A \cdot \overline{B} \cdot \overline{C}$.

$_C{}^{AB}$	00	01	11	10
0	1	1	1	1
1	0	1	0	0

Table 2.2 Truth tables and Karnaugh maps for the 5 basic gates

OR			NOR			AND			NAND			EXOR		
\multicolumn{15}{c}{**Truth tables**}														
A	B	F	A	B	F	A	B	F	A	B	F	A	B	F
0	0	0	0	0	1	0	0	0	0	0	1	0	0	0
0	1	1	0	1	0	0	1	0	0	1	1	0	1	1
1	0	1	1	0	0	1	0	0	1	0	1	1	0	1
1	1	1	1	1	0	1	1	1	1	1	0	1	1	0

Karnaugh maps:

$_B\!{}^A$	0	1	$_B\!{}^A$	0	1	$_B\!{}^A$	0	1	$_B\!{}^A$	0	1	$_B\!{}^A$	0	1
0	0	1	0	1	0	0	0	0	0	1	1	0	0	1
1	1	1	1	0	0	1	0	1	1	1	0	1	1	0

It is possible to lump variables together in groups of 2, 4 or 8—also with overlap. Here the four individual terms $\overline{A}\cdot\overline{B}\cdot\overline{C} + \overline{A}\cdot B\cdot\overline{C} + A\cdot B\cdot\overline{C} + \overline{A}\cdot B\cdot C$ have one common factor $\overline{C}$ and the last two terms $\overline{A}\cdot B\cdot\overline{C} + \overline{A}\cdot B\cdot C$ are all in the $\overline{A}\cdot B$ region.

Consequently, this gives the reduced function $F = \overline{A}\cdot B + \overline{C}$.

The same principle applies to a 4 variable Karnaugh maps:

CD\AB	00	01	11	10
00				
01				
11				
10				

Karnaugh maps are not just flat sheets. They are more akin to globes (like world maps) and can be rolled around. For example:

2.5 Karnaugh Maps

CD \ AB	00	01	11	10
00				
01	1			1
11	1			1
10				

Results in the function $F = \overline{B}\cdot D$ because where the 1s are B is always 0 and D is always 1.

Rule 24 from the rules of Boolean algebra.

$$\overline{A} + A \cdot B = \overline{A} + B.$$

This can be put into two Karnaugh maps:

$\overline{A} + A\cdot B$ $\overline{A} + B$ Which are identical.

B \ A	0	1
0	1	0
1	1	1

B \ A	0	1
0	1	0
1	1	1

Alternatively, using Boolean algebra:

$F = \overline{A} + A\cdot B \;=\; \overline{A} + B$

$\overline{F} = \overline{\overline{A} + A\cdot B} \;=\; \overline{\overline{A} + B}$

$\overline{F} = A\cdot(\overline{A} + \overline{B}) = A\cdot\overline{B}$

oder $F = \overline{A} + B$

Rule 25 from the rules of Boolean algebra.

$$A\cdot B + A\cdot C + \overline{B}\cdot C = A\cdot B + \overline{B}\cdot C$$

C \ AB	00	01	11	10
0	0	0	1	0
1	1	0	1	1

C \ AB	00	01	11	10
0	0	0	1	0
1	1	0	1	1

Taking the function depicted in the Karnaugh map below:

$_{CD}{}^{AB}$	00	01	11	10
00	1	1	1	1
01	1	0	0	1
11	1	0	0	1
10	1	1	1	1

The terms: $\overline{A}\cdot\overline{B} + A\cdot\overline{B} + \overline{C}\cdot\overline{D} + C\cdot\overline{D}$ can easily be identified. However instead of looking at the 1s we could also consider the 0s. This gives the inverse function $\overline{F} = B \cdot D$ or $F = \overline{B} + \overline{D}$. This is more compact and can be easily realised using one NAND-gate and an inverter.

After minimising a function, the next step is usually to produce the all NAND or all NOR version of the circuit. This allows the purchase of just one sort of device thus reducing costs and guaranteeing the same propagation delay for each gate. To achieve this, we must put the Boolean expression into Disjunctive or Conjunctive form.

2.6 Sum of Products and Product of Sums

In the initial stages, the first step in circuit design is minimization, i.e. utilizing the minimum number of devices. However, it is often better to use just one form of gate. An all NAND or all NOR design has two advantages:

1. Only one form of chip need be purchased—economies of scale!
2. Exactly the same propagation time delay for each identical device.

According to De Morgan's theory, inverting the inputs to a NOR-gate produces an AND-gate and inverting the inputs to a NAND-gate produces an OR-gate. By connecting both inputs together we can use either a NAND or NOR gate to produce in inverter.

For an all NAND solution the Boolean expression must be put into "Sum of Products" (SoP) form and for an all NOR solution into "Product of Sums" (PoS) form.

For example, with the EXOR function:

Sum of Products (SoP)—*disjunctive normal form* (DNF).

$$F_1 = A\cdot\overline{B} + \overline{A}\cdot B \qquad\qquad \text{EXOR } (A \oplus B)$$
$$\uparrow \quad\;\; \uparrow$$
$$\text{minterms (NAND)}$$

If a minterm $= 1$ then $F = 1$ (Rule 6).

$$F_1 = A\cdot\overline{B} + \overline{A}\cdot B \qquad \text{or} \qquad \overline{F_1} = \overline{A\cdot\overline{B} + \overline{A}\cdot B} \quad = \quad \overline{(A\cdot\overline{B})}\cdot\overline{(\overline{A}\cdot B)}$$
$$= (\overline{A} + B)\cdot(A + \overline{B}) \qquad \text{PoS form } (\overline{EXOR})$$

2.6 Sum of Products and Product of Sums

Product of Sums (PoS)—*conjunctive normal form* (CNF).

$$F_2 = (A + \overline{B}) \cdot (\overline{A} + B) \qquad \overline{\text{EXOR}\,(A \oplus B)}$$
$$\uparrow \qquad \uparrow$$
$$\text{maxterms (NOR)}$$

If a maxterm $= 0$ then $F = 0$ (Rule 8).

$$F_2 = (A + \overline{B}) \cdot (\overline{A} + B) \quad \text{or} \quad \overline{F_2} = \overline{(A + \overline{B}) \cdot (\overline{A} + B)} \quad = \quad \overline{A + \overline{B}} + \overline{\overline{A} + B}$$
$$= \overline{A} \cdot B + A \cdot \overline{B} \qquad \text{SoP form (EXOR)}$$

Which means $F_2 = \overline{F_1}$ or $F_1 = \overline{F_2}$.

Also, $F_2 = (A + \overline{B}) \cdot (\overline{A} + B) = A \cdot \overline{A} + A \cdot B + \overline{A} \cdot \overline{B} + B \cdot \overline{B}$

$$= A \cdot B + \overline{A} \cdot \overline{B} \qquad \overline{\text{EXOR}} \qquad \text{also disjunctive normal form.}$$

For example, the inverse function $\overline{F} = A \cdot \overline{B} \cdot \overline{C} + A \cdot B \cdot C$ can be put into a truth table in both SoP and PoS forms:

A B C	F	SoP	PoS
0 0 0	1	$\overline{A} \cdot \overline{B} \cdot \overline{C}$	
0 0 1	1	$\overline{A} \cdot \overline{B} \cdot C$	
0 1 0	1	$\overline{A} \cdot B \cdot \overline{C}$	
0 1 1	1	$\overline{A} \cdot B \cdot C$	
1 0 0	0		$A + \overline{B} + \overline{C}$
1 0 1	1	$A \cdot \overline{B} \cdot C$	
1 1 0	1	$A \cdot B \cdot \overline{C}$	
1 1 1	0		$\overline{A} + \overline{B} + \overline{C}$

Try and realize this function as a digital circuit!

Given the four combinations of two inputs to a gate, there are 16 possible outcomes as listed in Table 2.3. Some of these functions are well known, others are more obscure and not normally available in single TTL devices.

However, all these functions can be realised by using a single multiplexer. A multiplexer allows combinations of gates to be emulated with one single VLSI (Very Large Scale Integration) device. The basis of the multiplexer is a CMOS switch known as a transmission gate.

Table 2.3 The 16 possible functions which can be derived from two inputs

A	0	1	0	1	LSB	
B	0	0	1	1	MSB	
f_0	0	0	0	0	Ground (Logic 0)	
f_1	0	0	0	1	$A \cdot B$	AND
f_2	0	0	1	0	$\overline{A} \cdot B$	
f_3	0	0	1	1	B	
f_4	0	1	0	0	$A \cdot \overline{B}$	
f_5	0	1	0	1	A	
f_6	0	1	1	0	$A \oplus B$	EXOR
f_7	0	1	1	1	$A + B$	OR
f_8	1	0	0	0	$\overline{A + B}$	NOR
f_9	1	0	0	1	$\overline{A \oplus B}$	EXNOR
f_{10}	1	0	1	0	$\overline{A}$	
f_{11}	1	0	1	1	$\overline{A} + B$	
f_{12}	1	1	0	0	$\overline{B}$	
f_{13}	1	1	0	1	$A + \overline{B}$	
f_{14}	1	1	1	0	$\overline{A \cdot B}$	NAND
f_{15}	1	1	1	1	+Vcc (Logic 1)	

2.7 Transmission Gates and Tri-State

Transmission gates are bidirectional devices, so the IN and OUT terminals can be reversed, i.e. signals can be sent in both directions. Most transmission gates are capable of handling analogue as well as digital signals making them suitable for telephone systems, video routers, etc.

Referring to Fig. 2.5, with Logic 1 on A and Logic 0 on $\overline{A}$ both transistors conduct and pass the signal at IN to OUT. In the complementary condition, both transistors turn off forcing a high-impedance condition (tri-state) on both the IN and OUT nodes. Transmission gates such as the DS3690 can have many channels.

Tri-state is a condition which is neither logic 0 nor logic 1. When switched off, a tri-state output floats in a high impedance state. This means it can be connected to a bus system with impunity. Only when the device is enabled, does the output assume a distinct logic level.

2.8 Multiplexers

By combining transmission gates with a simple binary to decimal encoder (more of encoders later), multiplexers can be built. These can emulate any logic function. However, to design a circuit this function must be put into canonical form.

2.8 Multiplexers

Instead of logic functions generated by specific gates, it is possible to represent ALL logic functions with one device. This device is called a multiplexer (MUX in short).

In Fig. 2.6, as the select inputs S_1 and S_2 cycle through the usual binary sequence (00, 01, 10, 11), the transistors (or transmission gates) switch the inputs D_0, D_1, D_2 and D_3 to the output F. Although there are only 5 commonly used basic logic functions (AND, OR, NAND, NOR and EXOR), there are potentially 16 possibilities (Table 2.3).

Such a multiplexer works like a large 4 way switch. The binary values on S_1 and S_2 determine which input (D_0 to D_3) is selected and the result is put onto the output F. In most cases, the inputs D_0 to D_3 are hard wired depending on the logic function we

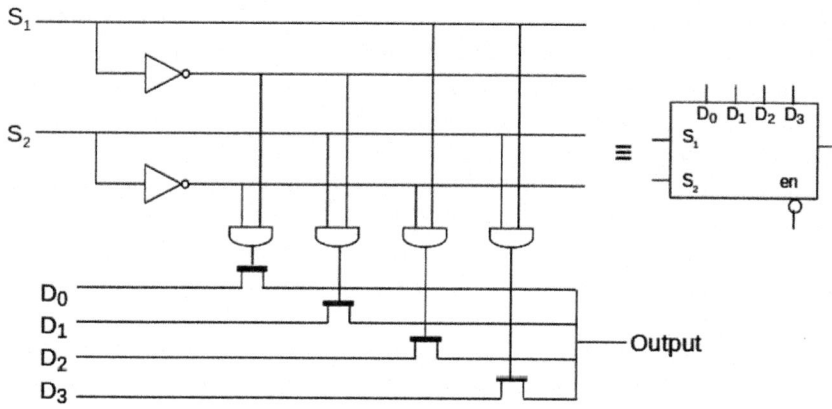

Fig. 2.6 4 bit Multiplexer

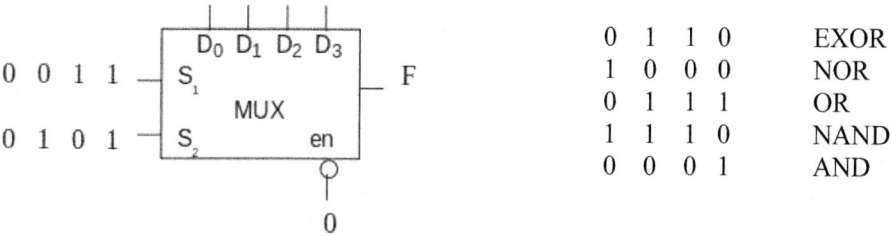

Fig. 2.7 Function of a 4-bit multiplexer

wish to emulate. However, these inputs can be controlled by other device outputs in more complicated designs.

Although extremely versatile, multiplexers are VLSI (very large-scale integration) chips and tend to be more expensive and slower than discrete devices. Typical examples of commercially available multiplexers are: 74HC153 (4 to 1) and 74HC151 (8 to 1).

2.9 Canonical Form

Hitherto the concentration has been on reducing logic functions to their minimum by eliminating redundancy. However, it is sometimes necessary to go in the opposite direction and deliberately introduce redundancy. The reason will soon become apparent.

Example
Take the function $F = \overline{A} + B \cdot \overline{C} + \overline{B} \cdot C$.

From rule 1 it is known that $(A + \overline{A})$ and $(B + \overline{B})$ and $(C + \overline{C})$ are all 1. If these terms are now included in F, then F remains effectively unchanged. Now each term in F contains A, B and C or their respective inverse.

Deliberately introducing redundancy gives:

$$F = \overline{A} \cdot (B + \overline{B}) \cdot (C + \overline{C}) + (A + \overline{A}) \cdot B \cdot \overline{C} + (A + \overline{A}) \cdot \overline{B} \cdot C.$$

Expanding F in the usual way yields:

$$F = \overline{A} \cdot B \cdot C + \overline{A} \cdot B \cdot \overline{C} + \overline{A} \cdot \overline{B} \cdot C + \overline{A} \cdot \overline{B} \cdot \overline{C} + A \cdot B \cdot \overline{C} + \overline{A} \cdot B \cdot \overline{C} + A \cdot \overline{B} \cdot C + \overline{A} \cdot \overline{B} \cdot C$$

Here it can be seen that some terms, namely $\overline{A} \cdot B \cdot \overline{C}$ and $\overline{A} \cdot \overline{B} \cdot C$ appear twice. These can be eliminated thus reducing F to:

$$F = \overline{A} \cdot B \cdot C + \overline{A} \cdot B \cdot \overline{C} + \overline{A} \cdot \overline{B} \cdot C + \overline{A} \cdot \overline{B} \cdot \overline{C} + A \cdot B \cdot \overline{C} + A \cdot \overline{B} \cdot C$$
$$0\,1\,1 \quad\;\; 0\,1\,0 \quad\;\; 0\,0\,1 \quad\;\; 0\,0\,0 \quad\;\; 1\,1\,0 \quad\;\; 1\,0\,1$$

This logic function can now put put directly into MUX form as shown in Fig. 2.8. The equivalent values in binary are written below the terms in F. These are the values which must be set to logic 1 on the $D_0 \ldots D_7$ inputs, the rest must be set to logic 0.

With an 8 bit multiplexer with three inputs (A, B, C) the values given in F are wired to the data inputs (D_0–D_7).

2.9 Canonical Form

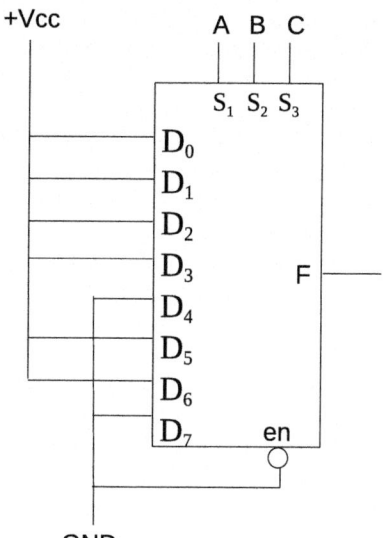

A B C	Selected	F	SoP	PoS
0 0 0	D₀	1	$\overline{A}\cdot\overline{B}\cdot\overline{C}$	
0 0 1	D₁	1	$\overline{A}\cdot\overline{B}\cdot C$	
0 1 0	D₂	1	$\overline{A}\cdot B\cdot\overline{C}$	
0 1 1	D₃	1	$\overline{A}\cdot B\cdot C$	
1 0 0	D₄	0		$A+\overline{B}+\overline{C}$
1 0 1	D₅	1	$A\cdot\overline{B}\cdot C$	
1 1 0	D₆	1	$A\cdot B\cdot\overline{C}$	
1 1 1	D₇	0		$A+B+C$

Fig. 2.8 8 bit multiplexer

By driving the inputs S_1, S_2 and S_3, the data bits on the inputs $D_0 \ldots D_7$ can be selected. In this example, only the binary D_4 and D_7 inputs corresponding to 1 0 0 and 1 1 1 result in $F = 0$. All the rest deliver $F = 1$. This can also be depicted using a Karnaugh map as shown below.

$S_1 S_2$ S_3	00	01	11	10
0	1	1	1	0
1	1	1	0	1

The inverted enable input allows the device to be switched on with a logic 0 and off with a logic 1. In the off state the output F is neither logic 1 nor logic 0 but the floating condition tri-state.

The table shows the equivalent Boolean conditions for disjunctive (SoP) and conjunctive (PoS) configurations, i.e.

$$F = \overline{A}\cdot\overline{B}\cdot\overline{C} + \overline{A}\cdot\overline{B}\cdot C + \overline{A}\cdot B\cdot\overline{C} + \overline{A}\cdot B\cdot C + A\cdot\overline{B}\cdot C + A\cdot B\cdot\overline{C} \quad \text{in SoP form}$$

and

$$F = (\overline{A}+\overline{B}+C) \cdot (A+\overline{B}+\overline{C}) \cdot (A+B+C) \quad \text{in PoS form}$$

Rule 20

This presents a good opportunity to prove Rule 20 which appears in PoS form:

$$(A + B) \cdot (B + C) \cdot (C + \overline{A}) = (A + B) \cdot (C + \overline{A})$$

$$\underset{B \text{ (rule 3)}}{} \qquad \underset{0 \text{ (rule 4)}}{}$$

$$(A \cdot B + A \cdot C + B \cdot B + B \cdot C) \cdot (C + \overline{A}) = A \cdot C + A \cdot \overline{A} + B \cdot C + B \cdot \overline{A}$$

$$\underset{0 \text{ (rule 4)} \quad C \text{ (rule 3)} \quad 0 \text{ (rule 4)}}{}$$

$$A \cdot B \cdot C + A \cdot B \cdot \overline{A} + A \cdot C \cdot C + A \cdot C \cdot \overline{A} + B \cdot C + B \cdot \overline{A} + B \cdot C + B \cdot C \cdot \overline{A} = A \cdot C + B \cdot C + B \cdot \overline{A}$$

one of the B·C terms is superfuous

$$A \cdot B \cdot C + A \cdot C + B \cdot \overline{A} + B \cdot C + B \cdot C \cdot \overline{A} = A \cdot C + B \cdot C + B \cdot \overline{A}$$

$$\underset{1 \text{ (rules 1 \& 6)}}{}$$

$$B \cdot C \cdot (A + 1 + \overline{A}) + A \cdot C + B \cdot \overline{A} = A \cdot C + B \cdot C + B \cdot \overline{A}$$

$$B \cdot C + A \cdot C + B \cdot \overline{A} = A \cdot C + B \cdot C + B \cdot \overline{A}$$

Rule 23

Similarly, the proof of Rule 23 which is in SoP form:

$$A \cdot B + B \cdot C + \overline{A} \cdot C = A \cdot B + \overline{A} \cdot C$$

Putting both sides into canonical form:

$$A \cdot B \cdot (C + \overline{C}) + (A + \overline{A}) \cdot B \cdot C + \overline{A} \cdot (B + \overline{B}) \cdot C = A \cdot B \cdot (C + \overline{C}) + \overline{A} \cdot (B + \overline{B}) \cdot C$$

Expanding:

$$A \cdot B \cdot C + A \cdot B \cdot \overline{C} + A \cdot B \cdot C + \overline{A} \cdot B \cdot C + \overline{A} \cdot B \cdot C + \overline{A} \cdot \overline{B} \cdot C = A \cdot B \cdot C + A \cdot B \cdot \overline{C} + \overline{A} \cdot B \cdot C + \overline{A} \cdot \overline{B} \cdot C$$

On the left hand side the terms A·B·C and $\overline{A}$·B·C appear twice so eliminating redundancy gives:

$$A \cdot B \cdot C + A \cdot B \cdot \overline{C} + \overline{A} \cdot B \cdot C + \overline{A} \cdot \overline{B} \cdot C = A \cdot B \cdot C + A \cdot B \cdot \overline{C} + \overline{A} \cdot B \cdot C + \overline{A} \cdot \overline{B} \cdot C$$

i.e. both sides of the equation are identical.

Taking this a step further, larger combinations of gates may be replaced by programmable logic. For the programming of such systems a special Boolean logic programming language VHDL has been developed. With VHDL the *behaviour* of an *entity* with the required *architecture* can be defined.

2.10 Programmable Logic Devices

The outputs of a combinational system depend only on the state of the system inputs. Therefore, such a system can be replaced by a properly programmed memory. Every combination of inputs connected to the memory address bus, select a cell to be programmed so as to deliver the proper value to the data lines, which are connected to the outputs.

In addition, it should be pointed out that every sequential system is made of a combinational subsystem with several outputs looping back to be used as inputs. Hence, even a sequential system may also be replaced by a properly programmed memory.

Consequently, every system based on logic gates (wired logic) may be replaced by a system based on a programmed memory (programmable logic) resulting in an equivalent behaviour. Programmable Logic Devices (PLDs) are systems based on this philosophy.

2.11 PROM (Programmable Read-Only Memory)

The structure of a Programmable Read-Only Memory (PROM), showing the contents of the decoder driving the selection lines, can be represented as shown in Fig. 2.9.

A horizontal line is being activated if all the corresponding vertical address lines are high (see **AND** gates). An output line is being activated if one of the connected horizontals is high (see **OR** gates). The other programmable logic devices are based on a similar structure.

Fig. 2.9 PROM

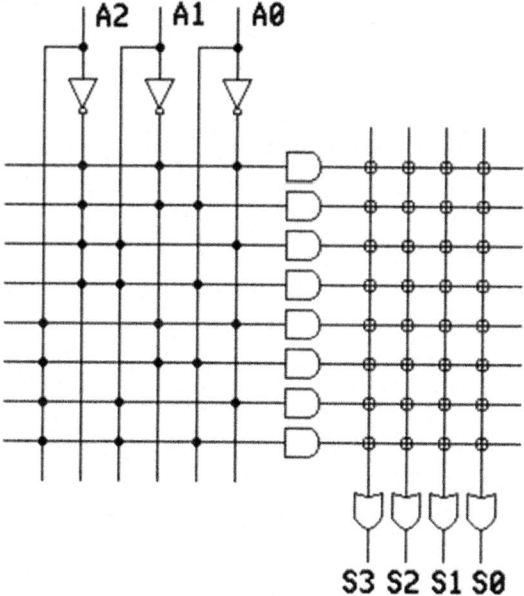

Fig. 2.10 PAL

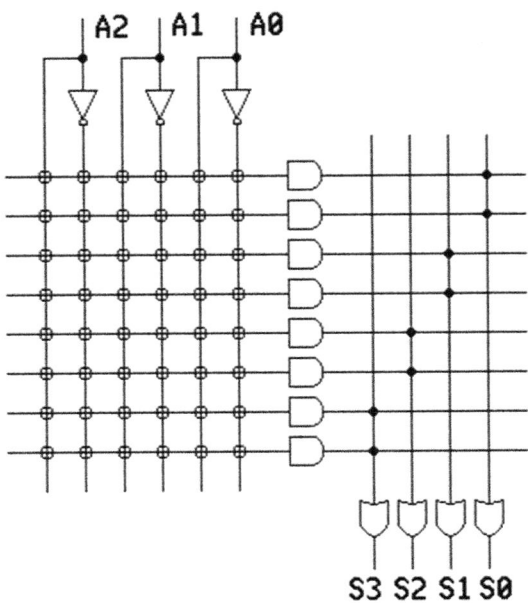

2.12 PAL (Programmable Array Logic)

PAL devices use the same principles as described above, but with a programmable connection matrix on the left side (**AND** gates), and pre-determined connections on the right side (**OR** gates), as in the following example (Fig. 2.10).

Programming of PALs is in Sum of Products (disjunctive) form.

2.13 PLA (Programmable Logic Array)

PLAs are devices where the two connection matrices (**AND** side and **OR** side) are both programmable, also in SOP form (Fig. 2.11).

Component implementation is facilitated and simplified because there are less chips. Connection routing is made easier for design and production, since it is easier to route one bus than many individual lines. Maintenance is improved, since it is much easier and faster to program a PLA than to redesign a fixed circuit implementation.

As an example, the 74S330 programmable chip (fuse technology) integrates 12 input lines, 50 **AND** lines and 6 output lines, into a 20 pin package (12 inputs, 6 outputs, 2 power supply lines).

Fig. 2.11 PLA

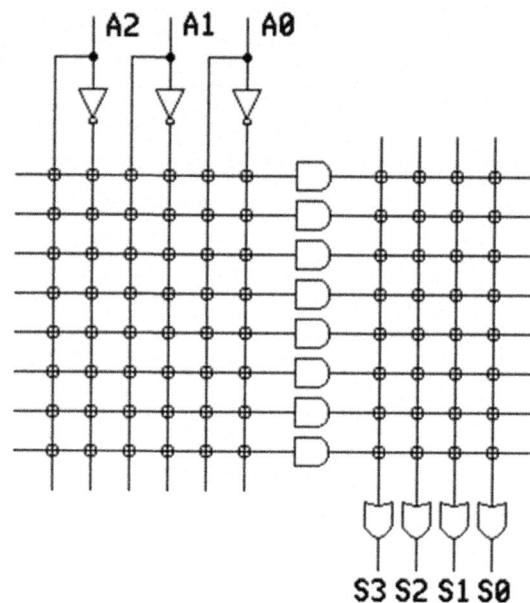

2.14 GAL (Generic Array Logic)

Simple PLA and PAL devices are gradually being replaced by Generic Array Logic (GAL). These comprise programmable AND gate inputs and a range of programmable macro cells for sequential logic functions. Generic Array Logic (GAL) devices are an improvement on the PAL concept because one device type is able to take the place of many PAL device types or can even have functionality not covered by the original range of PAL devices. Its primary benefit, however, is that they are erasable and re-programmable, making prototyping and design changes easier.

2.15 FPGA (Field Programmable Gate Arrays)

Field Programmable Gate Arrays (FPGA) can be produced using a number of Configurable Logic Blocks (CLBs). FPGA devices contain typically between tens of thousands to several million gates. More flexible logic functions than sum-of-product expressions are provided. These include complicated feedback paths between macro cells and specialized logic for implementing various commonly used functions, such as integer arithmetic [4].

2.16 CPLD (Complex Programmable Logic Device)

A complex programmable logic device (CPLD) is a programmable logic device with complexity between that of PALs and FPGAs and architectural features of both. The main building block of the CPLD is a macrocell, which contains logic implementing disjunctive normal form expressions and more specialized logic operations. CPLDs typically have the equivalent of thousands to tens of thousands of logic gates, while PALs typically have a few hundred gate equivalents at the most.

The most noticeable difference between a large CPLD and a small FPGA is the presence of on-chip non-volatile memory in the CPLD, which allows CPLDs to be used for "boot loader" functions, before handing over control to other devices not having their own permanent program storage. A good example is where a CPLD is used to load configuration data for an FPGA from non-volatile memory.

The original CPLD devices included analogue sense amplifiers to boost the performance. However, this performance boost came at the cost of higher current requirements. Most modern CPLD are purely digital. However, mixed analog and digital devices are also a current area of interest.

2.17 Example VHDL Code for OR and AND Functions

Very High Density Logic (VHDL) is a computer language designed for programming programmable devices. The following two examples represent simple logic functions hitherto covered in this book.

2.17 Example VHDL Code for OR and AND Functions

OR function

```
-------------------------------------
--OR Gate
-------------------------------------

library ieee;
use ieee.std_logic_1164.all;

-------------------------------------

entity OR_ent is
port(   x: in std_logic;
        y: in std_logic;
        F: out std_logic
);
end OR_ent;

-------------------------------------

architecture OR_arch of OR_ent is
begin

    process(x, y)
    begin
        -- compare to truth table
        if ((x='0') and (y='0')) then
            F <= '0';
        else
            F <= '1';
        end if;
    end process;

end OR_arch;

architecture OR_beh of OR_ent is
begin

    F <= x or y;

end OR_beh;
```

AND function

```
---------------------------------------
-- AND Gate
---------------------------------------

library ieee;
use ieee.std_logic_1164.all;

---------------------------------------

entity AND_ent is
port(   x: in std_logic;
        y: in std_logic;
        F: out std_logic
);
end AND_ent;
```

```
architecture behav1 of AND_ent is
begin

    process(x, y)
    begin
        -- compare to truth table
        if ((x='1') and (y='1')) then
            F <= '1';
        else
            F <= '0';
        end if;
    end process;

end behav1;

architecture behav2 of AND_ent is
begin

    F <= x and y;

end behav2;
```

As can be seen, for anyone with programming experience, the function is fairly obvious. <http://esd.cs.ucr.edu/labs/tutorial/>

2.18 Binary Hardware Addition and Subtraction

Not only is it possible to carry out Boolean operations but we can add binary values together. The mathematical function of addition of binary variables may be achieved using a particular combination of gates. The first building block is the so called half adder. The half adder produces a SUM (S) through the XOR function and a carry (C) by means of an AND function on the inputs A and B (Fig. 2.12).

For full addition, two half adders must be joined together with an additional OR gate (Fig. 2.13).

The truth table for a half adder is:

A	B	C=A·B	S=A⊕B	Decimal
0	0	0	0	0
0	1	0	1	1
1	0	0	1	1
1	1	1	0	2

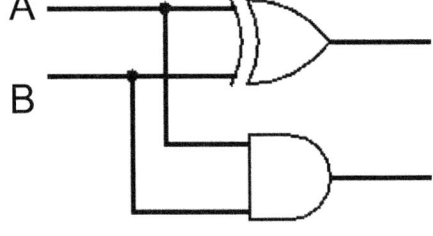

Fig. 2.12 Half adder circuit and truth table

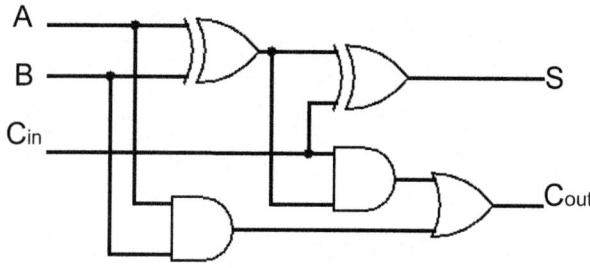

Fig. 2.13 Full adder circuit

Together with the basic Boolean functions of AND, OR and NOT, a very simple arithmetic and logic unit (ALU) can be built.

There are also subtractors which work in the same way, except one of the inputs to the AND gate is inverted. This results in a *Difference* and *Borrow* instead of a *Sum* and *Carry*.

2.19 Subtractor

Combining an Exclusive-OR gate with a NAND gate results in a simple digital binary Half Subtractor which produces sum (difference) and borrow bits for the next stage, as illustrated in Fig. 2.14.

Just as the concatenation of two half adders yields a full adder, the same is true for subtractors (Fig. 2.15).

As with the binary adder, it is also possible to have a number of 1-bit full binary subtractors connected or "cascaded" together to subtract two parallel n-bit numbers from each other. For example two 4-bit binary numbers.

The only difference between a full adder and a full subtractor is the inversion of one of the inputs. By using inversion, the process of subtraction becomes an addition with two's complement on all the bits in the subtrahend and setting the carry input of the least significant bit to logic "1" (Fig. 2.16).

The 74LS83, 74LS283 or CD4008 are 4-bit full-adder/full-subtractor circuits with a single control input for selecting between the two operations.

However, in real computers subtractors are rarely used. Instead *complement* and *addition* are the usual means of achieving subtraction (more of this in Binary mathematics).

A further useful function is comparison. Comparators are devices capable of deciding whether a value is equal to, less than or greater than another value. Given simple addition and a comparator, more complicated processor structures can be built.

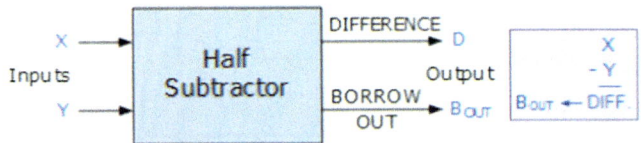

$$D = X \oplus Y \quad B = \overline{X} . Y$$

Fig. 2.14 Half subtractor circuit (Courtesy: Electronic tutorials)

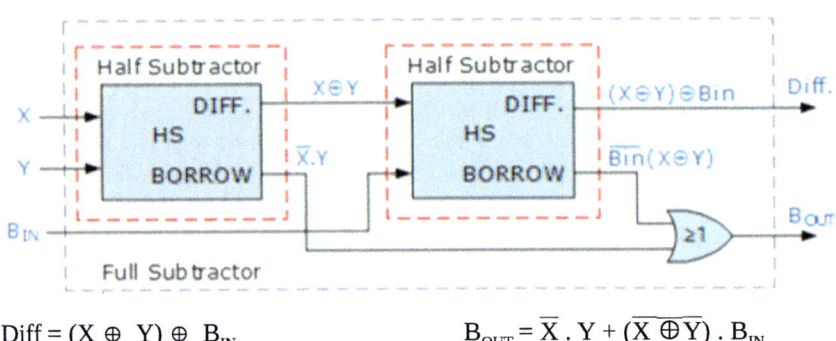

$\text{Diff} = (X \oplus Y) \oplus B_{IN}$ $\qquad B_{OUT} = \overline{X} . Y + (\overline{X \oplus Y}) . B_{IN}$

Fig. 2.15 Full subtractor circuit (Courtesy: Electronic tutorials)

2.20 Comparators

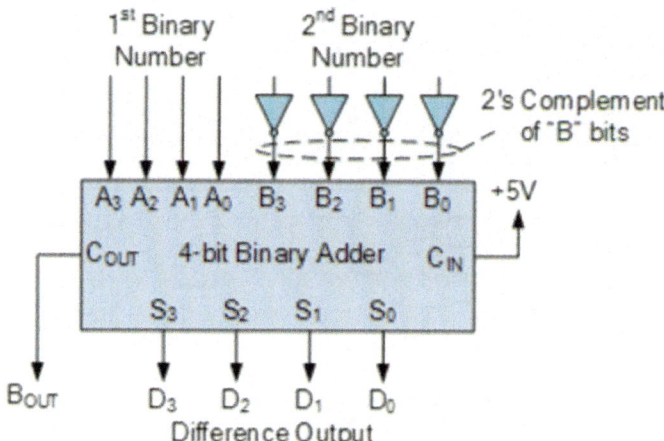

Fig. 2.16 Full adder/subtractor device (Courtesy: Electronic tutorials)

2.20 Comparators

Comparators enable two binary values to be compared. This is important because it must be decided whether a value is less than zero (i.e. negative) to make subtraction possible (Fig. 2.17).

$$\text{Greater} = A1 \cdot \overline{B1} + A0 \cdot \overline{B1} \cdot \overline{B0} + A1 \cdot A0 \cdot \overline{B0}$$
$$\text{Same} = \overline{GR + KL}$$
$$\text{Smaller} = B1 \cdot \overline{A1} + \overline{A1} \cdot \overline{A0} \cdot B0 + \overline{A0} \cdot B1 \cdot B0$$

Later it will be shown how microprocessors make branching decisions based on the comparison of values in a sequential program.

The data provided for such operations must be converted from a form we understand (ASCII keyboard symbols) into Binary. ASCII is the standard American Standard Code for Information Interchange which has been used since the existence of the first computer interfaces. These were usually in the form of modified typewriters with an electrical interface similar to those used in teletype terminals. In Table 2.4, the first 32 decimal values are so called "non-printing" ASCII characters. These were used as commands for the terminal or printer. Today printers generate such commands internally dependant on the text to be printed.

In a computer, the decimal values representing text symbols must be converted to binary (or the other way around). In Table 2.4 they are given in Hex and Octal so the

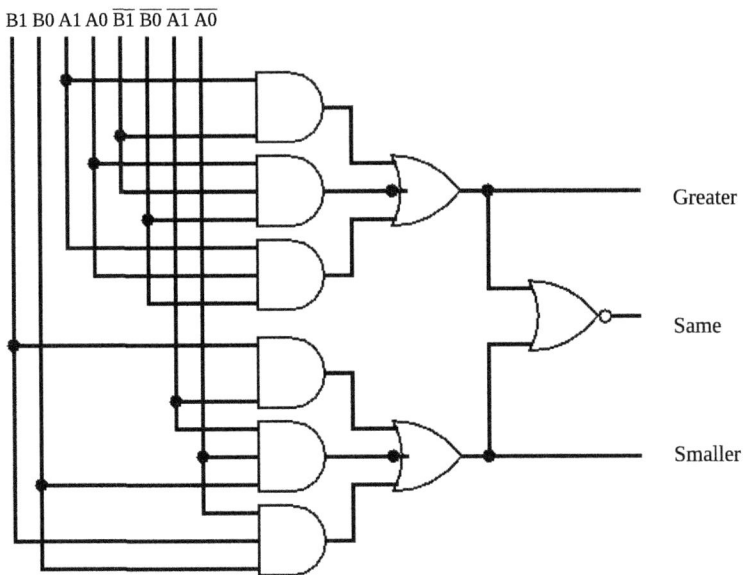

Fig. 2.17 Comparator

step to binary is trivial (see Chap. 4—Binary Maths). For data output, another often used conversion is from BCD to 7-segment display format.

2.21 Encoders and Decoders

As mentioned above, it is often necessary to convert binary into decimal and decimal to binary. The truth table below shows the 4 decimal possibilities (D_0, D_1, D_2 and D_3) and equivalent 2 bit binary representation (A and B).

D_0	D_1	D_2	D_3	A	B
1	0	0	0	0	0
0	1	0	0	0	1
0	0	1	0	1	0
0	0	0	1	1	1

2.21 Encoders and Decoders

Table 2.4 ASCII symbols and values (Coutesy: Lookup Tables.com)

```
Dec Hx Oct Char                        Dec Hx Oct Html   Chr   Dec Hx Oct Html Chr   Dec Hx Oct Html  Chr
 0  0 000 NUL (null)                    32 20 040 &#32; Space  64 40 100 &#64;  @    96 60 140 &#96;   `
 1  1 001 SOH (start of heading)        33 21 041 &#33;  !    65 41 101 &#65;  A    97 61 141 &#97;   a
 2  2 002 STX (start of text)           34 22 042 &#34;  "    66 42 102 &#66;  B    98 62 142 &#98;   b
 3  3 003 ETX (end of text)             35 23 043 &#35;  #    67 43 103 &#67;  C    99 63 143 &#99;   c
 4  4 004 EOT (end of transmission)     36 24 044 &#36;  $    68 44 104 &#68;  D   100 64 144 &#100;  d
 5  5 005 ENQ (enquiry)                 37 25 045 &#37;  %    69 45 105 &#69;  E   101 65 145 &#101;  e
 6  6 006 ACK (acknowledge)             38 26 046 &#38;  &    70 46 106 &#70;  F   102 66 146 &#102;  f
 7  7 007 BEL (bell)                    39 27 047 &#39;  '    71 47 107 &#71;  G   103 67 147 &#103;  g
 8  8 010 BS  (backspace)               40 28 050 &#40;  (    72 48 110 &#72;  H   104 68 150 &#104;  h
 9  9 011 TAB (horizontal tab)          41 29 051 &#41;  )    73 49 111 &#73;  I   105 69 151 &#105;  i
10  A 012 LF  (NL line feed, new line)  42 2A 052 &#42;  *    74 4A 112 &#74;  J   106 6A 152 &#106;  j
11  B 013 VT  (vertical tab)            43 2B 053 &#43;  +    75 4B 113 &#75;  K   107 6B 153 &#107;  k
12  C 014 FF  (NP form feed, new page)  44 2C 054 &#44;  ,    76 4C 114 &#76;  L   108 6C 154 &#108;  l
13  D 015 CR  (carriage return)         45 2D 055 &#45;  -    77 4D 115 &#77;  M   109 6D 155 &#109;  m
14  E 016 SO  (shift out)               46 2E 056 &#46;  .    78 4E 116 &#78;  N   110 6E 156 &#110;  n
15  F 017 SI  (shift in)                47 2F 057 &#47;  /    79 4F 117 &#79;  O   111 6F 157 &#111;  o
16 10 020 DLE (data link escape)        48 30 060 &#48;  0    80 50 120 &#80;  P   112 70 160 &#112;  p
17 11 021 DC1 (device control 1)        49 31 061 &#49;  1    81 51 121 &#81;  Q   113 71 161 &#113;  q
18 12 022 DC2 (device control 2)        50 32 062 &#50;  2    82 52 122 &#82;  R   114 72 162 &#114;  r
19 13 023 DC3 (device control 3)        51 33 063 &#51;  3    83 53 123 &#83;  S   115 73 163 &#115;  s
20 14 024 DC4 (device control 4)        52 34 064 &#52;  4    84 54 124 &#84;  T   116 74 164 &#116;  t
21 15 025 NAK (negative acknowledge)    53 35 065 &#53;  5    85 55 125 &#85;  U   117 75 165 &#117;  u
22 16 026 SYN (synchronous idle)        54 36 066 &#54;  6    86 56 126 &#86;  V   118 76 166 &#118;  v
23 17 027 ETB (end of trans. block)     55 37 067 &#55;  7    87 57 127 &#87;  W   119 77 167 &#119;  w
24 18 030 CAN (cancel)                  56 38 070 &#56;  8    88 58 130 &#88;  X   120 78 170 &#120;  x
25 19 031 EM  (end of medium)           57 39 071 &#57;  9    89 59 131 &#89;  Y   121 79 171 &#121;  y
26 1A 032 SUB (substitute)              58 3A 072 &#58;  :    90 5A 132 &#90;  Z   122 7A 172 &#122;  z
27 1B 033 ESC (escape)                  59 3B 073 &#59;  ;    91 5B 133 &#91;  [   123 7B 173 &#123;  {
28 1C 034 FS  (file separator)          60 3C 074 &#60;  <    92 5C 134 &#92;  \   124 7C 174 &#124;  |
29 1D 035 GS  (group separator)         61 3D 075 &#61;  =    93 5D 135 &#93;  ]   125 7D 175 &#125;  }
30 1E 036 RS  (record separator)        62 3E 076 &#62;  >    94 5E 136 &#94;  ^   126 7E 176 &#126;  ~
31 1F 037 US  (unit separator)          63 3F 077 &#63;  ?    95 5F 137 &#95;  _   127 7F 177 &#127;  DEL
```
Source: www.LookupTables.com

From decimal to binary its very simple as can be seen from the encoder circuit in Fig. 2.18. Here only simple diodes are needed to prevent unwanted current from flowing back onto the data lines. Only one data line may be selected at any one time. The binary equivalent is given on the outputs A and B.

To go from Binary to decimal, the A and B inputs must be converted to the decimal (D_0, D_1, D_2 and D_3) outputs.

A	B	D_0	D_1	D_2	D_3
0	0	1	0	0	0
0	1	0	1	0	0
1	0	0	0	1	0
1	1	0	0	0	1

For this some AND gates are needed, as shown in the decoder circuit of Fig. 2.19.

4 bit binary goes from 0000 to 1111 which is $2^4 = 16$ combinations. However, only 10 of them can be used to represent decimal values. Limiting the amount of combinations to only include the first 10 values (0000 to 1001) which represent only the decimal numbers is called BCD (Binary Coded Decimal). Of course such devices are available in single chips. For example: 8 to 3 line 74,148 encoder and BCD-Decimal 7442 decoder.

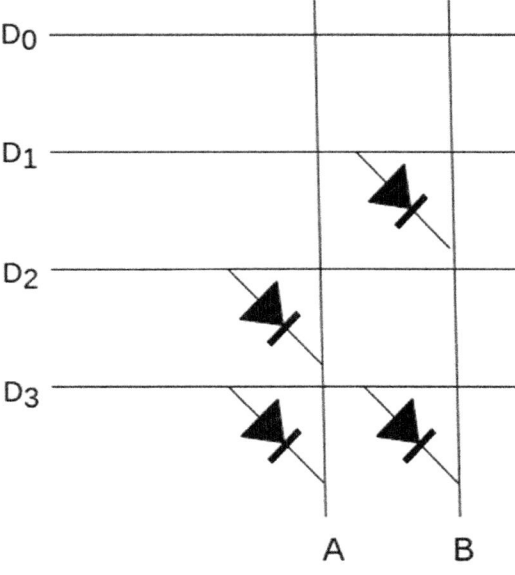

Fig. 2.18 Decimal to binary conversion

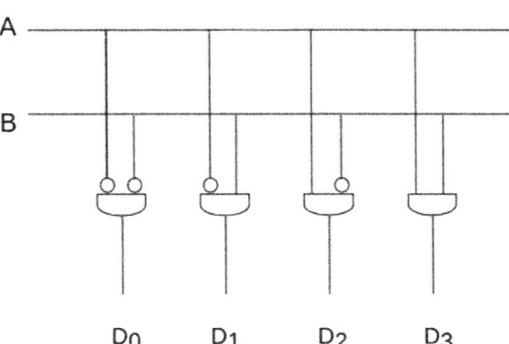

Fig. 2.19 Binary to decimal conversion

There are many different codes used in data acquisition and manipulation. XS3 (excess 3) is BCD plus 3. In this code the sequences 0000 and 1111 are missing. This is deliberate because XS3 was devised as a code which avoids the use of states which could be mistaken as short circuit or open circuit conditions in parallel bus lines.

In order to drive 7-segment displays, a special BCD to 7-segment decoder is needed.

2.22 Arithmetic and Logic Units (ALU)

The essential building blocks for the internal "brain" of a computer have now been considered. These will now be put together to form what is known as an arithmetic and logic unit (ALU).

2.22 Arithmetic and Logic Units (ALU)

Most computers contain only one circuit to perform AND, OR and the addition of two machine words. For n-bit operations, n identical circuits are connected in parallel. The full adder is one of the main elements of such a circuit as can be seen in a very simple arithmetic and logic unit (ALU) of Fig. 2.20. Depending on the binary values on the inputs F_0 and F_1 (00, 01, 10 or 11), such a circuit can perform A AND B, A OR B, $\overline{B}$ or A + B operations.

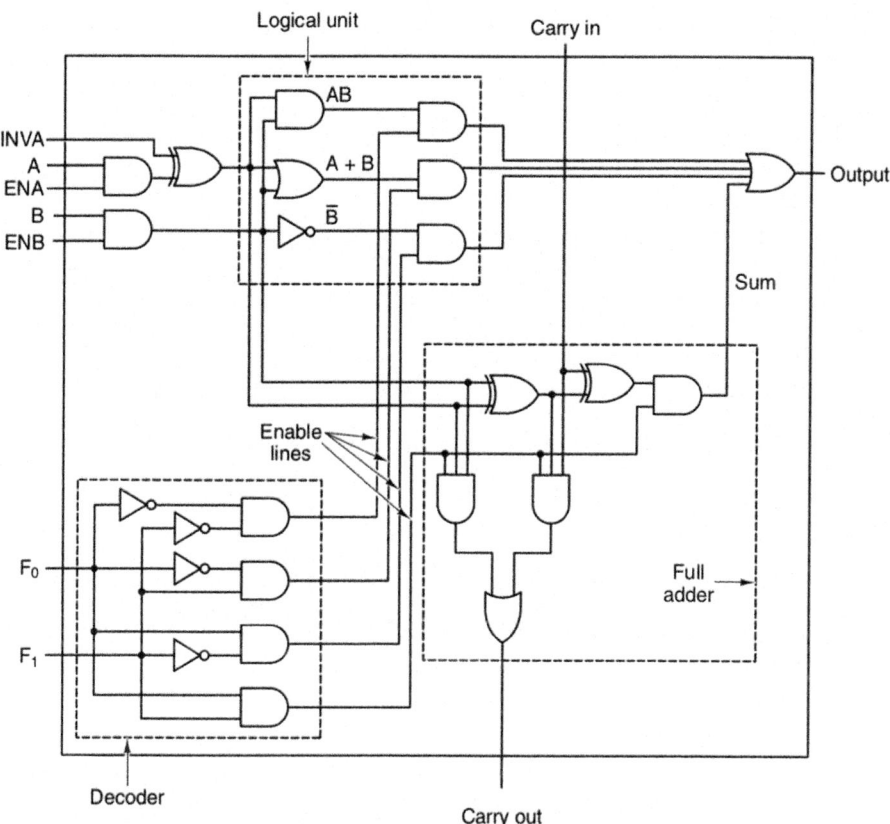

Fig. 2.20 1-bit ALU with functions AND, OR, NOT and addition [5]

The various functions of the ALU can be selected by means of the two inputs F_0 and F_1 as shown in the truth table below:

F_0	F_1	A	B	C	D	Function
0	0	1	0	0	0	A·B
0	1	0	1	0	0	A+B
1	0	0	0	1	0	$\overline{B}$
1	1	0	0	0	1	Sum

The circuit shown in Fig. 2.20 is highly simplified in comparison to a real ALU used in a modern computer but should help the reader to understand the basic principles. More complicated processors will be dealt with in Chap. 5.

2.23 Digital Simulators

Simulation of circuits may be performed by suitable software. Similar software is also available for programming industrial PLCs (Programmable Logic Controllers). There are a number of logic simulators on the internet (usually free) to download.

LogiSim <http://www.cburch.com/logisim/> runs under Java: /DT/Logisim3—right click on logisim-generic-2.7.1.jar and open with "Oracle Java 7 runtime" or higher.

GEDA and SPICE <https://graahnul-grom.github.io/ref-docs/geda-csygas.html> also for Linux.

CEDAR <http://sourceforge.net/projects/cedarlogic/> for Microsoft Windows.

Tetzl LogicSim <http://www.tetzl.de/java_logic_simulator.html> runs under Java.

For a more extensive list of simulation systems, the reader should consult <https://en.wikipedia.org/wiki/List_of_free_electronics_circuit_simulators>.

Later it will be shown how microcontrollers and microprocessors can be used to make industrial computers (PLCs). There are many PLC systems (Allen Bradley, Beckhoff, Omron, Siemens, Wago, etc.). Most come with a software package which includes some form of simulation. This allows the development and simulation of logic circuits without the PLC hardware.

Appendix—Logic Families

Transistor-Transistor Logic (TTL)
Standard 74-series. Typical propagation delay 9 nS, typical power consumption 10 mW. Rarely used today.

Schottky
Faster 74S-series.

Low-Power Schottky
74LS-series. Typical propagation delay 7 nS, typical power consumption 2 mW.

Advanced Schottky
74AS-series. Typical propagation delay 1.5 nS, typical power consumption 2 mW.

Advanced Low-Power Schottky
74ALS-series. Typical propagation delay 4 nS, typical power consumption 1 mW.

GHz TTL
74G-Series. First TTL over 1 GHz using 3 V logic.

CMOS
4000 (5 to 15 Volt capability) and 74C-series. Typical propagation delay 30 nS, typical static power consumption 400 µW.

High Speed CMOS
74HC-series. Typical propagation delay 18 nS, typical static power consumption 400 µW. 4000-series compatible.

74HCT-series. Typical propagation delay 18 nS, typical static power consumption 400 µW. TTL compatible.

Advanced CMOS
74AC-series. Typical propagation delay 3.3 nS for 5 Volt operation and 5 nS for 3.3 Volt operation, typical static power consumption 200 µW.

Advanced High-Speed CMOS
74AHC-series. Typical propagation delay 5.4 nS for 5 Volt operation and 8.3 nS for 3.3 Volt operation, typical static power consumption 200 µW. 4000-series compatible.

74AHCT-series. Typical propagation delay 5.4 nS for 5 Volt operation and 8.3 nS for 3.3 Volt operation, typical static power consumption 200 μW. TTL compatible.

74FCT-series. Very fast TTL-compatible.

Low-Voltage CMOS
74LV-series. Typical propagation delay 6.5 nS both for 5 Volt and 3.3 Volt operation, typical static power consumption 100 μW.

74LVC-series. 74LV series with symmetrical outputs.

74ALVC-series. Faster Version of 74LVC.

74FCT-T-series. 74FCT 2.4 Volt variation with TTL-compatible outputs.

Advanced Low-Voltage CMOS
74LV-series. Typical propagation delay 3 nS both for 5 Volt and 3.3 Volt operation, typical static power consumption 200 μW.

Microgates
74AC1G-series. Single gate configurations commonly used in laptop computers.

BICMOS 74BCT-series. Lower power consumption at high frequency. Propagation delay between 4.8 and 7 nS.

Advanced BICMOS
74ABT-series. Designed for bus-interface functions. Propagation delay between 2.9 and 4.6 nS.

Advanced Low-Power BICMOS
74ALB-series. Designed for bus-interface functions. Maximum propagation delay of 2.2 nS.

Low-Voltage BICMOS
74LVT-series. 3.3 Volt version of the 74ALB-series. Maximum propagation delay of 4.6 nS.

Emitter-Coupled
ECL-devices. Very high speed (>500 MHz) bit slice operation. Propagation delay less than 1 nS. Typical power consumption 60 mW per gate.

Appendix—Logic Families

Datasheet and Pin Out Sources

https://www.findchips.com/

https://search.datasheetcatalog.net/

https://www.alldatasheet.com/

https://www.ti.com/

http://www.unitechelectronics.com/7400-7448data.htm

http://en.wikipedia.org/wiki/List_of_7400_series_integrated_circuits

http://www.qsl.net/yo5ofh/data_sheets/data_sheets_page.htm.

Exercises 2—Combinational Logic

Exercise 2.1

Simplify the Boolean expression: $F = \overline{A} \cdot (B + \overline{C}) \cdot (A + \overline{B} + C) \cdot \overline{A} \cdot \overline{B} \cdot \overline{C}$.
Sketch the corresponding circuit diagram and truth table.

$F = \overline{A} \cdot (B + \overline{C}) \cdot (A + \overline{B} + C) \cdot \overline{A} \cdot \overline{B} \cdot \overline{C}$

multiply out:

$F = \overline{A} \cdot (A \cdot B + B \cdot \overline{B} + B \cdot C + A \cdot \overline{C} + \overline{B} \cdot \overline{C} + C \cdot \overline{C}) \cdot \overline{A} \cdot \overline{B} \cdot \overline{C}$

$B \cdot \overline{B}$ and $C \cdot \overline{C}$ disappear (Rule 4)

$= (\overline{A} \cdot A \cdot B + \overline{A} \cdot B \cdot C + \overline{A} \cdot A \cdot \overline{C} + \overline{A} \cdot \overline{B} \cdot \overline{C}) \cdot \overline{A} \cdot \overline{B} \cdot \overline{C}$

$= (\overline{A} \cdot B \cdot C \cdot \overline{A} \cdot \overline{B} \cdot \overline{C} + \overline{A} \cdot \overline{B} \cdot \overline{C} \cdot \overline{A} \cdot \overline{B} \cdot \overline{C}) = \overline{A} \cdot \overline{B} \cdot \overline{C}$

Look at Exercise2-1.circ for the simulation of the $(\overline{A} \cdot B \cdot C + \overline{A} \cdot \overline{B} \cdot \overline{C})$ part.

A B C	F	
0 0 0	1	$= \overline{A} \cdot \overline{B} \cdot \overline{C}$
0 0 1	0	
0 1 0	0	
0 1 1	1	$= (\overline{A} \cdot B \cdot C)$
1 0 0	0	
1 0 1	0	
1 1 0	0	
1 1 1	0	

Exercise 2.2

Simplify the Boolean expression: $F = (A + B) \cdot (\overline{A \cdot B} + C) + A \cdot B$

$F = (A + B) \cdot (\overline{A \cdot B} + C) + A \cdot B$

From rule 22: $\overline{A \cdot B} = \overline{A} + \overline{B}$

$F = (A + B) \cdot (\overline{A} + \overline{B} + C) + A \cdot B$

Expanding gives:

$F = A \cdot \overline{A} + A \cdot \overline{B} + A \cdot C + \overline{A} \cdot B + B \cdot \overline{B} + B \cdot C + A \cdot B$

$F = A \cdot \overline{B} + A \cdot C + \overline{A} \cdot B + B \cdot C + A \cdot B$

$F = A \cdot (B + \overline{B}) + A \cdot C + \overline{A} \cdot B + B \cdot C$

From rule 1: $(B + \overline{B}) = 1$

$F = A \cdot (1 + C) + \overline{A} \cdot B + B \cdot C$

From rule 6: $(1 + C) = 1$

$F = A + \overline{A} \cdot B + B \cdot C$

From rule 18: $A + \overline{A} \cdot B = A + B$

$F = A + B \cdot (1 + C)$

$\underline{F = A + B}$

Appendix—Logic Families

Exercise 2.3

Design a circuit for the function: $\overline{A} \cdot B + A \cdot \overline{B}$ using only NAND-Gates.

Design a circuit for the function: $\overline{A} \cdot B + A \cdot \overline{B}$ using only NOR-Gates.

....

$\overline{A} \cdot B + A \cdot \overline{B}$ with only NAND-Gates.
The function is already in SOP-Form—Exercide 2.3a. circ.
$\overline{A} \cdot B + A \cdot \overline{B}$ with only NOR-Gates.
$F = \overline{A} \cdot B + A \cdot \overline{B}$.
$F = \overline{\overline{A} \cdot B \cdot A \cdot \overline{B}}$.
$F = (\overline{\overline{A}} + \overline{B}) \cdot (\overline{A} + \overline{\overline{B}})$.
$F = (A + \overline{B}) \cdot (\overline{A} + B)$.
Now in POS-Form—Exercise 2.3b.circ.

Exercise 2.4

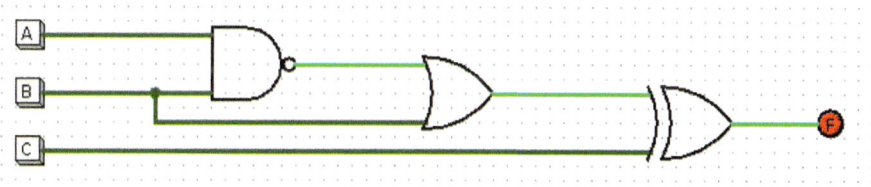

Minimize the above circuit with the help of truth tables, Karnaugh maps and Boolean algebra.

....

The LogiSim circuit is also given in Exercise 2.4.circ.
Truth table:

A B C	F
0 0 0	1
0 0 1	0
0 1 0	1
0 1 1	0
1 0 0	1

Karnaugh map:

C\AB	00	01	11	10
0	1	1	1	1
1	0	0	0	0

Boolean Algebra:

$F = (\overline{A.B} + B) \oplus C$

$F = (\overline{A.B} + B) \oplus \cdot \overline{C} + (\overline{A.B} + B) \oplus \cdot C$ $\qquad x \oplus y = x \cdot \overline{y} + \overline{x} \cdot y$

$F = (\overline{A} + \overline{B} + B) \cdot \overline{C} + A \cdot B \cdot \overline{B} \cdot C$

$F = (\overline{A} + 1) \cdot \overline{C}$

$F = \overline{C}$

Exercise 2.5

Put the function $F = A \cdot B \cdot C \cdot D + \overline{A} \cdot B \cdot C \cdot D + A \cdot \overline{C} \cdot D + A \cdot \overline{C} \cdot \overline{D} + \overline{A} \cdot B \cdot \overline{C}$ into a Karnaugh map and reduce F to three terms.

··

$F = A \cdot B \cdot C \cdot D + \overline{A} \cdot B \cdot C \cdot D + A \cdot \overline{C} \cdot D + A \cdot \overline{C} \cdot \overline{D} + \overline{A} \cdot B \cdot \overline{C}$
$\qquad\ \ $ 1111$\quad$ 0111$\quad$ 1 0 1$\quad$ 1 0 0$\ \ $ 0 1 0

CD\AB	00	01	11	10
00	0	1	1	1
01	0	1	1	1
11	0	1	1	0
10	0	0	0	0

CD\AB	00	01	11	10
00	0	1	1	1
01	0	1	1	1
11	0	1	1	0
10	0	0	0	0

$F = B \cdot \overline{C} + A \cdot \overline{B} \cdot \overline{C} + B \cdot C \cdot D$

Appendix—Logic Families

Exercise 2.6

Put the function $\bar{B} + B \cdot \bar{D}$ into a Karnaugh map, a truth table and a logic circuit.

$F = \bar{B} + B \cdot \bar{D}$

CD\AB	00	01	11	10
00	1	1	1	1
01	1	0	0	1
11	1	0	0	1
10	1	1	1	1

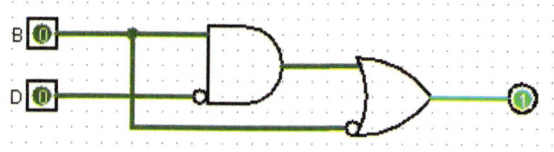

Oder $F = \bar{B} + \bar{D}$.

D\B	0	1
0	1	1
1	1	0

An alternative would be the inverse function:

CD\AB	00	01	11	10
00	1	1	1	1
01	1	0	0	1
11	1	0	0	1
10	1	1	1	1

$F = \overline{B \cdot D}$

Boolean algebra: $F = \bar{B} + B \cdot \bar{D}$

$\bar{F} = \overline{\bar{B} + B \cdot \bar{D}} = \bar{\bar{B}} + \overline{B \cdot \bar{D}} = B \cdot (\bar{B} + \bar{\bar{D}}) = B \cdot (\bar{B} + D) = B \cdot D$

$\therefore F = \overline{B \cdot D}$

Exercise 2.7

The following four Functions depict the XOR function:

(a) $F = \overline{A} \cdot B + A \cdot \overline{B}$ (b) $F = (A + \overline{B}) \cdot (\overline{A} + B)$ (c) $F = A \cdot B + \overline{A} \cdot \overline{B}$ (d) $F = (A + B) \cdot (\overline{A} + \overline{B})$

Sketch the circuits for each function. Which are in conjunctive and which is disjunctive form?

....

$F1 = \overline{A} \cdot B + A \cdot \overline{B}$ SOP (disjunctive)
$\overline{F2} = (A + \overline{B} \cdot (\overline{A} + B)$ POS (conjunctive)
$\overline{F3} = A \cdot B + \overline{A} \cdot \overline{B}$ SOP (disjunctive)
$F4 = (A + B) \cdot (\overline{A} + \overline{B})$ POS (conjunctive)

Simulation in Exercise 2.7.circ.

Exercise 2.8

Put the following function into canonical Form and realize using a Multiplexer.

$$F = A \cdot \overline{B} + A \cdot \overline{C} + A \cdot B + B \cdot \overline{C} + A \cdot C + \overline{B} \cdot C$$

Appendix—Logic Families

Design the same circuit using discrete logic elements.

$$F = A \cdot \overline{B} + A \cdot \overline{C} + A \cdot B + B \cdot \overline{C} + A \cdot C + \overline{B} \cdot C$$

Minimizing: $F = A \cdot (\overline{B} + \overline{C} + B + C) + B \cdot \overline{C} + \overline{B} \cdot C = A + B \cdot \overline{C} + \overline{B} \cdot C$

Expanding in canonical Form:

$$F = A \cdot B \cdot C + A \cdot \overline{B} \cdot C + A \cdot B \cdot \overline{C} + A \cdot \overline{B} \cdot \overline{C} + A \cdot B \cdot \overline{C} + \overline{A} \cdot B \cdot \overline{C} + A \cdot \overline{B} \cdot C + \overline{A} \cdot \overline{B} \cdot C$$
$$\quad 1\,1\,1 \quad 1\,0\,1 \quad 1\,1\,0 \quad 1\,0\,0 \quad 1\,1\,0 \quad 0\,1\,0 \quad 1\,0\,1 \quad 0\,0\,1$$

Remove redundancy:

$$F = A \cdot B \cdot C + A \cdot \overline{B} \cdot C + A \cdot B \cdot \overline{C} + A \cdot \overline{B} \cdot \overline{C} + \overline{A} \cdot B \cdot \overline{C} + \overline{A} \cdot \overline{B} \cdot C$$
$$\quad 1\,1\,1 \quad 1\,0\,1 \quad 1\,1\,0 \quad 1\,0\,0 \quad 0\,1\,0 \quad 0\,0\,1$$

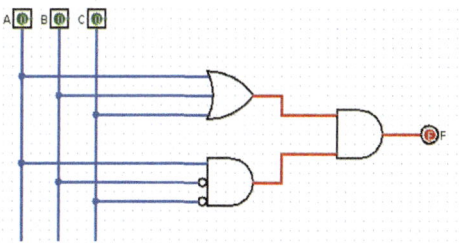

A B C	F	
0 0 0	0	$\overline{F}$
0 0 1	1	
0 1 0	1	
0 1 1	0	$\overline{F}$
1 0 0	1	
1 0 1	1	
1 1 0	1	
1 1 1	1	

$F = A + B \cdot \overline{C} + \overline{B} \cdot C \quad$ oder $\quad \overline{F} = \overline{A + B \cdot \overline{C} + \overline{B} \cdot C}$

$\overline{F} = \overline{A} \cdot \overline{B \cdot \overline{C}} \cdot \overline{\overline{B} \cdot C} = \overline{A} \cdot (\overline{B} + C) \cdot (B + \overline{C}) = \overline{A} \cdot (B \cdot \overline{B} + B \cdot C + \overline{B} \cdot \overline{C} + C \cdot \overline{C})$

$\overline{F} = \overline{A} \cdot \overline{B} \cdot \overline{C} + \overline{A} \cdot B \cdot C \quad$ in SOP-Form
$\quad\quad 0\,0\,0 \quad\quad 0\,1\,1$

and further in POS-Form:

$$F = \overline{\overline{A} \cdot \overline{B} \cdot \overline{C} + \overline{A} \cdot B \cdot C.}$$
$$= \overline{\overline{A} \cdot \overline{B} \cdot \overline{C}} \cdot \overline{\overline{A} \cdot B \cdot C.}$$
$$= (A + B + C) \cdot (A + \overline{B} + \overline{C}).$$

Exercise 2.9
Put the sum and carry functions of a full adder in canonical form. Realize this with two multiplexers. Sketch the Karnaugh maps for this.

....

$C_o = A \cdot B + A \cdot C_i + B \cdot C_i$

in canonical Form:

$C_o = A \cdot B \cdot C_i + A \cdot B \cdot \overline{C_i} + A \cdot \overline{B} \cdot C_i + \overline{A} \cdot B \cdot C_i$
$ + \overline{A} \cdot B \cdot C_i$

$C_o = A \cdot B \cdot C_i + A \cdot B \cdot \overline{C_i} + A \cdot \overline{B} \cdot C_i + \overline{A} \cdot B \cdot C_i$
 1 1 1 1 1 0 1 0 1 0 1 1

$S = \overline{A} \cdot \overline{B} \cdot C_i + \overline{A} \cdot B \cdot \overline{C_i} + A \cdot B \cdot C_i + A \cdot \overline{B} \cdot \overline{C_i}$
 0 0 1 0 1 0 1 1 1 1 0 0

as Karnaugh map:
C_o

C_i\AB	00	01	11	10
0	0	0	1	0
1	0	1	1	1

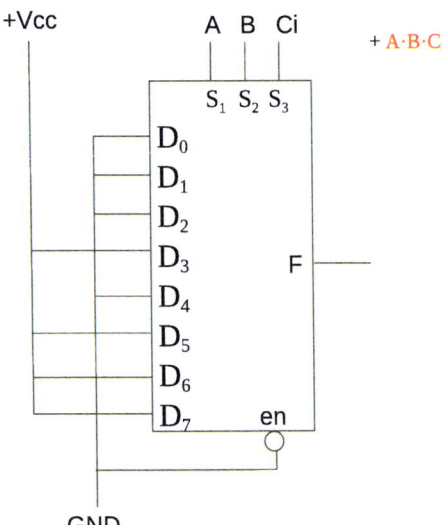

Appendix—Logic Families

S

c_i \ AB	0 0	0 1	1 1	1 0
0	0	1	0	1
1	1	0	1	0

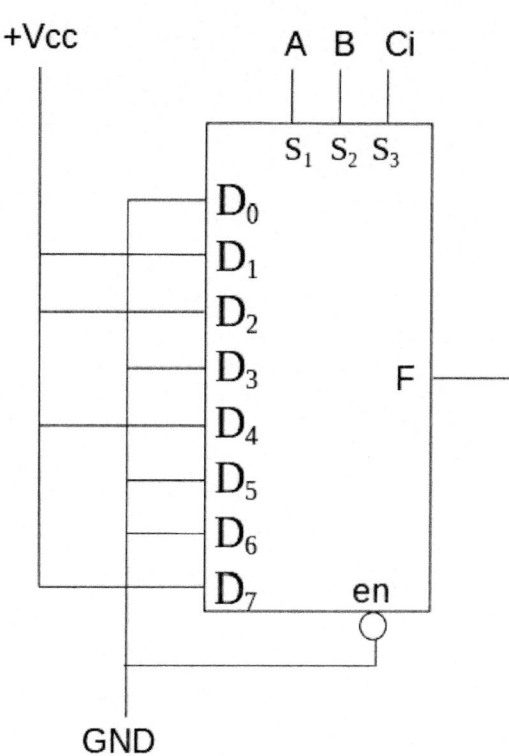

Exercise 2.10
Derive a Karnaugh map for a 2-bit comparator.

···

$B_0 B_1$ \ $A_0 A_1$	00	01	11	10
0 0	A=B	A>B	A>B	A>B
0 1	A<B	A=B	A>B	A<B
1 1	A<B	A<B	A=B	A<B
1 0	A<B	A>B	A>B	A=B

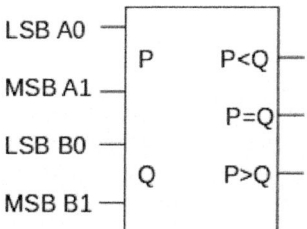

Exercise 2.11
Create a truth table to drive a 7-segment display from a 4-digit BCD value.

....

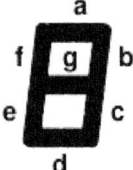

d	BCD	a	b	c	d	e	f	g
0	0000	1	1	1	1	1	1	0
1	0001	0	1	1	0	0	0	0
2	0010	1	1	0	1	1	0	1
3	0011	1	1	1	1	0	0	1
4	0100	0	1	1	0	0	1	1
5	0101	1	0	1	1	0	1	1
6	0110	0	0	1	1	1	1	1
7	0111	1	1	1	0	0	0	0
8	1000	1	1	1	1	1	1	1
9	1001	1	1	1	1	0	1	1

References

1. Maddock R.J. & D.M. Calcutt—*Electronics for Engineers*—Longman UK, 1994.
2. Green. D.C.—*Applied Digital Electronics*—Longman, 1999.
3. Boole. G.—The mathematical analysis of logic, being an essay towards a calculus of deductive reasoning.—MacMillan, London 1847.
4. Seals. R.C & G.F. Whapshott—Programmable Logic: PLDs and FPGAs—Springer, 1997.
5. Tanenbaum A.S & T. Austin—Structured Computer Organization—Pearson, 2013. ISBN 10: 0-13-291652-5; ISBN 13: 978-0-13-291652-3.
6. Dietmayer D.L.—Logic Design of Digital Systems—Allyn & Bacon, 1988.

References

7. Gosling P.E. & Q.L.M. Laarhoven—*CODES for Computers and Microprocessors*—MacMillan, 1980.
8. Maddock. R.J. & D.M. Calcutt—Electronics for Engineers—Longman Scientific & Technical—1. Juni 1994 {ISBN: 0582215838, 9780582215832}.
9. Neidel. M & P. Schrader—*Mikrocontroller und GAL*—Pflaum, München, 1997. ISBN: 3-7-905-0752-0.
10. Parr. E.A.—*Logic designers handbook*—Butterworth-Heinemann, 1993.
11. Schubert. M,—Mixed Analog-Digital Signal Modeling Using Event-Driven VHDL—*SBCCI*, Gramado, Brazil, 25-27 Aug. 1997.
12. Texas Instruments—The TTL Data Book for Design Engineers.—1984.
13. Wakerly. J.F.—*Digital Design*—Prentice Hall, 1994.

Internet Sites

14. http://www.6502.org/users/dieter/mt15b/mt15b_1.htm
15. http://www.electronics-tutorials.ws/combination/binary-subtractor.html
16. https://learn.sparkfun.com/tutorials/logic-levels/ttl-logic-levels
17. https://www.lookuptables.com/text/ascii-table
18. https://tinytapeout.com/runs/tt02/081/
19. https://www.xilinx.com/products/silicon-devices/cpld/cpld.html

Sequential Logic 3

Sequential logic differs from simple combinational logic in that feedback paths may be included. This introduces much greater time dependency, as will be seen as the text progresses. Nevertheless, only two binary states are allowed and the same rules of boolean algebra apply. The basic building blocks of sequential logic circuits are bistable devices of which there are many historical examples.

The simplest form of bistable (or Flip-flop) can be made from 2 inverting logic gates. Previously such memories were realised mechanically or using vacuum tubes. In fact almost any two active devices can be configured as a memory cell. With the advent of transistors and logic gates, more practical devices could be designed such as the SR (Set/Reset) Flip-Flop. This forms the basic modern memory element.

3.1 SR Flip Flop

The SR (Set—Reset) flip-flop is the simplest bistable circuit of all and easiest to understand. It is basically a device which has two inputs and two complementary outputs. A pulse on one of the inputs causes an output to take on a particular logical state. The outputs will then remain in this state until a similar pulse is applied to the other input. The two inputs are called the Set and Reset (sometimes called Preset and Clear). Such flip-flops can be made simply by cross coupling two inverting gates. Either NAND or NOR gates can be used (Fig. 3.1).

To describe the circuit using NAND gates, assume that initially both S and R are at the logic 1 state and that the Q output is at the logic 0 state.

Fig. 3.1 SR Flip-Flop using two cross-coupled NAND gates

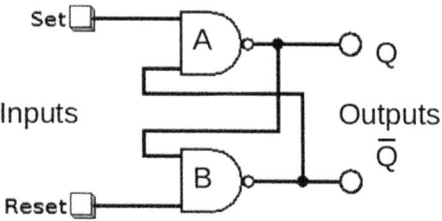

Now, if Q = 0 and Reset = 1, then these are the states of the inputs to gate B, therefore the output of gate B is 1 (making $\overline{Q}$ = 1). The output of gate B is connected to an input of gate A so if Set = 1, both inputs of gate A are logic 1. This means that the output of gate A must be 0 (as was originally specified). In other words, the 0 state at Q is continuously disabling gate B so that any change on the Reset input has no effect. Also the 1 state at Q is continuously enabling gate A so that any change to the Set input will be transmitted through to Q. The above conditions constitute one of the stable states of the device referred to as the Reset state since Q = 0. At this point, should both Set = 1 and Reset = 1 then the Q output will remain at 0 and $\overline{Q}$ at 1 (the last output state will be maintained—memory).

Now suppose that with the SR flip-flop in the Reset state, the Set input goes to 0. The output of gate A (Q) = 1 and with Reset = 1, the output of gate B ($\overline{Q}$) will go to 0. With $\overline{Q}$ now 0, gate A is disabled maintaining Q at 1. Consequently, when the Set input returns to the 1 state it has no effect on the flip-flop whereas a change at the Reset input will cause a change in the output of gate B. The above conditions constitute the other stable state of the device, called the Set state since Q = 1. Note that the change in the state of the Set input from 1 to 0 has caused the flip-flop to change from the Reset state to the Set state. At this point, should both Set = 1 and Reset = 1 then the Q output will remain at 1 and $\overline{Q}$ at 0 (the last output state will be maintained—memory).

There is another input condition which has not yet been considered. That is when both the Reset and Set inputs are taken to logic 0. When this happens both Q and $\overline{Q}$ will be forced to 1 and will remain so far as long as Set and Reset inputs are maintained at 0. However when both inputs return to 1 there is no way of knowing whether the flip-flop will go into the Reset state or the Set state. The condition is said to be indeterminate. Because of this indeterminate state great care must be taken when using SR flip-flops to ensure that both inputs are not set to 0 simultaneously (Fig. 3.2).

The alternative configuration comprises two cross coupled NOR gates. In this circuit, if Set = 1 then the initial state of $\overline{Q}$ = 0 (and consequently Q = 1) will be maintained as long as Reset = 0. Should Set change to 0 then the flip-flop remembers the last state (memory). Should Set = 0 and Reset = 1 then Q changes to 0 and $\overline{Q}$ to 1. The state Set = 1 and Reset = 1 results in an indeterminate output state.

3.2 Latches

Fig. 3.2 SR Flip-Flop using two cross-coupled NOR gates

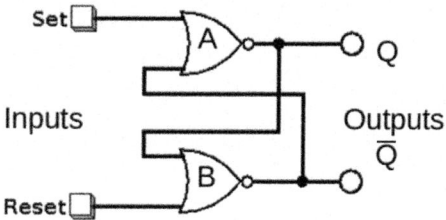

These flip-flops can be simulated using Logisim written by Carl Burch. Simply run "logisim-generic-2.7.1.jar" under Oracle Java 7 or 8 Runtime (or higher). See Chap. 2 for more information.

3.2 Latches

Latches differ from flip-flops in that they are level controlled. Whereas flip-flops are triggered either by a leading or falling edge of a clock pulse, latches are driven by the logic level (high or low) of the clock input. They are a common alternative to clocked flip-flops in many applications.

In fact the very first simple flip-flops considered (without clock inputs) were really latches. A good example is the 74LS279. As shown in Fig. 3.3, latches can be active at logic 1 levels or active at logic 0 levels (Fig. 3.4).

Fig. 3.3 Active high (logic 1) or active low (logic 0) latches

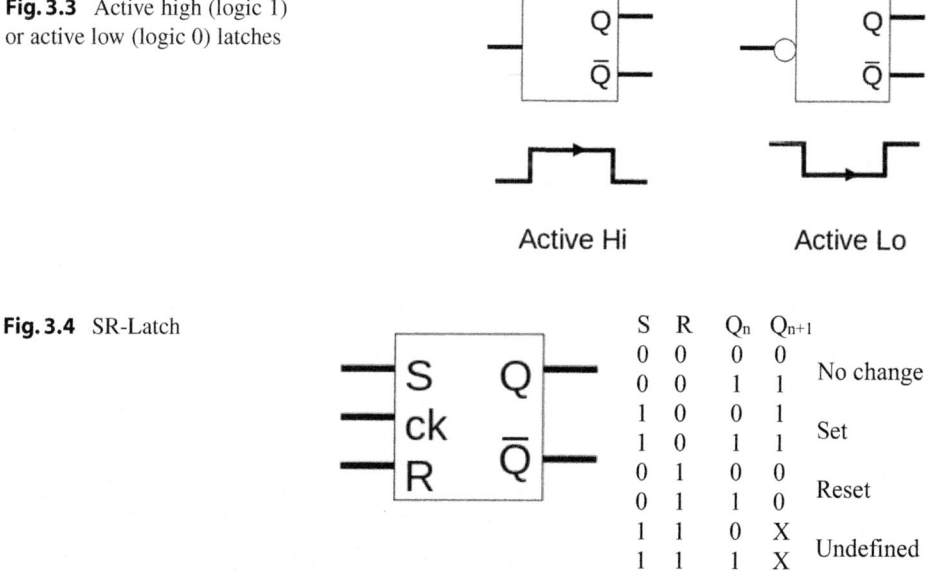

Fig. 3.4 SR-Latch

Clock	Q_{n+1}
Lo	Q_n
Hi	D

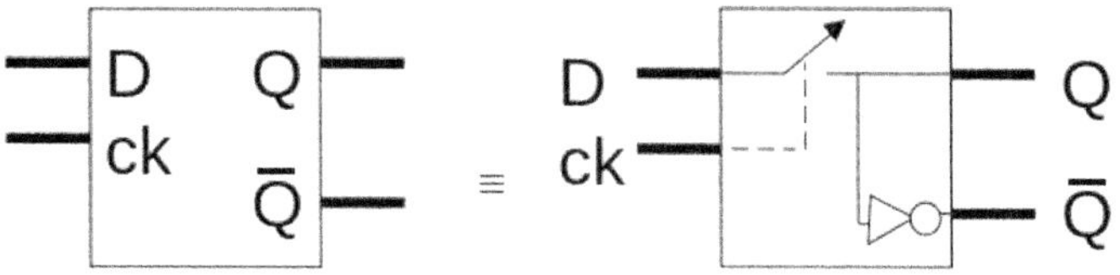

Fig. 3.5 D-Latch

Another variation is the D-Latch. This acts as a switch similar to a transmission gate in that whatever is on the D-input will be transferred to the Q-output as long as the clock input is high, otherwise the Q-output remains unchanged (Fig. 3.5).

Latches are often used in conjunction with counters where all the outputs must be given out simultaneously after they have reached a stable condition. For example Gray code to binary converters. A more advanced package, containing multiple tri-state D-latches, is the 74AHC373. There are also toggle flip-flops and latches. All flip-flops and latches can also be described by means of state diagrams.

3.2.1 The D-type Flip-Flop

The D type flip-flop (Data flip-flop) works like a switch which allows data to pass (or not) from the D-input to the Q-output. In Fig. 3.6 two different methods for converting SR flip-flops to a D-type are shown. The above circuit is the traditional gated D-type configuration with an additional inverter. The lower circuit works exactly the same but without the inverter, thus saving a gate. As with all flip-flop configurations, D-type flip-flops can be implemented with NAND or NOR gates with or without the additional set and clear.

The use of an inverter between the inputs ensures that the S and R inputs are always a complement of each other, thus eliminating the undefined condition of $S = R = 1$. As a result, the D flip-flop is also called a "transparent latch" because the output Q follows the D input when the clock input is high (CLK $= 1$) thus transferring the binary information at the input directly to the output as if the flip-flop were not present—effectively making it transparent.

3.2.2 The T Flip-Flop (Toggle Flip-Flop)

The toggle flip-flop, shown in Fig. 3.7, changes its state when the leading edge of the clock input is applied and remains unchanged after the clock pulse falling edge. Only when the next rising edge arrives does the output switch again, which gives the flip-flop

3.2 Latches

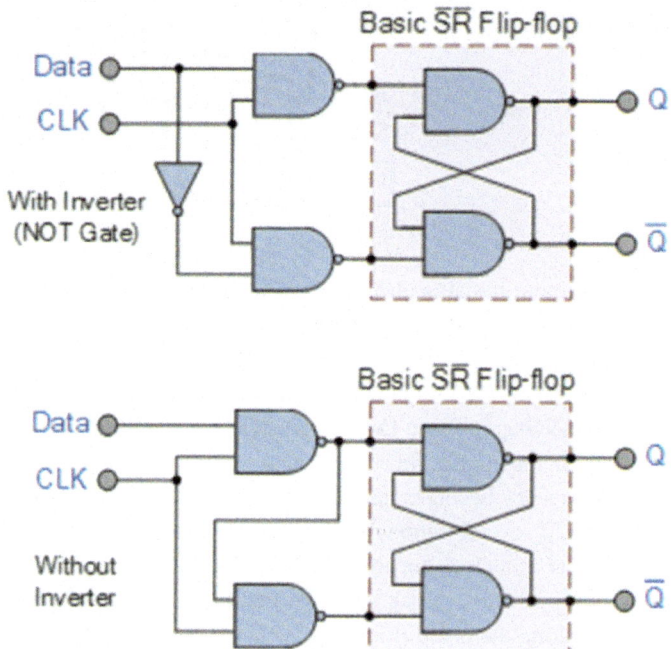

Fig. 3.6 D-Flip-Flops (Courtesy: Electronics-tutorials)

its name. The T-type (toggle) flip-flop is the basic building block of many digital circuits such as frequency dividers and digital counters.

JK flip-flops can be converted to T flip-flops in two simple ways. In the first variation, the J and K inputs can be connected to high as shown, with the clock input becoming the toggle. The second option is to connect the J and K inputs together to provide the toggle

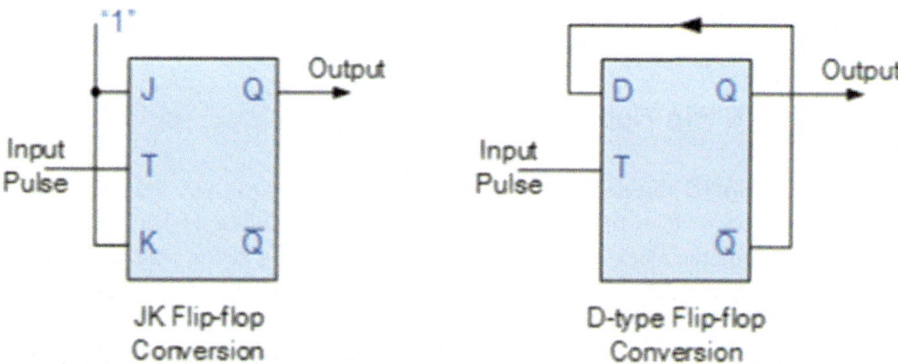

Fig. 3.7 T Flip-Flops (Courtesy: Electronics-tutorials)

Fig. 3.8 D Flip-Flops configured as T Flip-Flops in a frequency divider

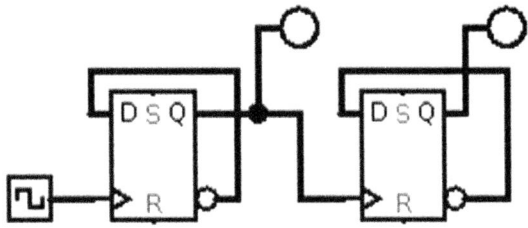

input while the clock input remains unchanged. The output toggles when T and CLK = 1. The output remains unchanged when T or CLK are low.

The D flip-flop can be converted into a T flip-flop just like the JK flip-flop by connecting the Q output directly to the D input, with the toggle signal T being the clock input, as shown above. Connecting $\overline{Q}$ to the D-input creates the required feedback.

Since the output of the toggle flip-flop changes state every time a clock signal is applied, its output frequency becomes half the frequency of the input signal and thus acts as a frequency divider. If several toggle flip-flops are cascaded to form a chain, the output of the first flip-flop acts as a clock for the second T flip-flop in the cascade arrangement, and the second flip-flop acts as a clock input for the third T flip-flop and so on, creating a frequency division along the chain (Fig. 3.8).

Flip-flops and latches form the basic building blocks of sequential logic circuits. Therefore, many IC manufacturers produce a variety of different types flip-flop chips using both TTL and CMOS technology such as: 74LS73, 74LS107, 74C109, 74C112 etc.

Complete computer memory systems are produced by connecting simple D flip-flops in a 2D array as shown in Fig. 3.9.

Flip-flops are the basic building blocks of modern computer memory. However, bistable (two stable state) devices have a long history and can be made from almost any pair of active elements. It is interesting to observe the historical development of memory from purely mechanical methods to the modern semiconductors.

A more sophisticated, and probably the most commonly used bistable device, is the JK flip-flop.

3.3 The JK Flip Flop

The gated SR NAND flip flop suffers from two basic problems: one, the S = 0 and R = 0 conditions (S = R = 0) must always be avoided. In the case of latches, if S or R change state while the clock input is high the correct latching action may not occur. To overcome these two fundamental design problems with the SR flip-flop, the JK flip-flop was developed.

The JK flip-flop is basically a gated SR flip-flop with the addition of clock input circuitry that prevents the illegal or invalid output condition which can occur when both

3.3 The JK Flip Flop

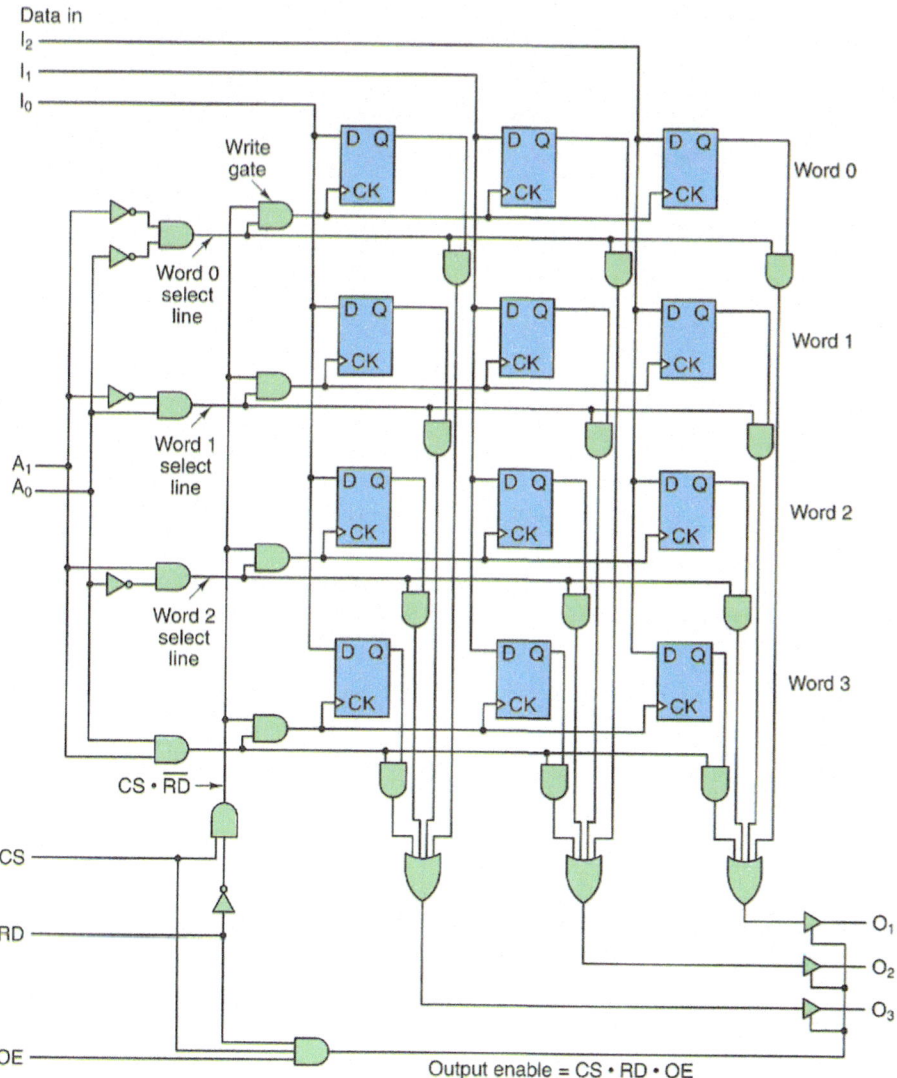

Fig. 3.9 Memory Matrix (Courtesy: E. Prabakar)

inputs S and R are equal to logic 1. Due to this additional clocked input, a JK flip-flop has four possible input combinations, "logic 1″, "logic 0″, "no change" and "toggle". The symbol for a JK flip-flop is similar to that of an SR bistable latch with the addition of a clock input.

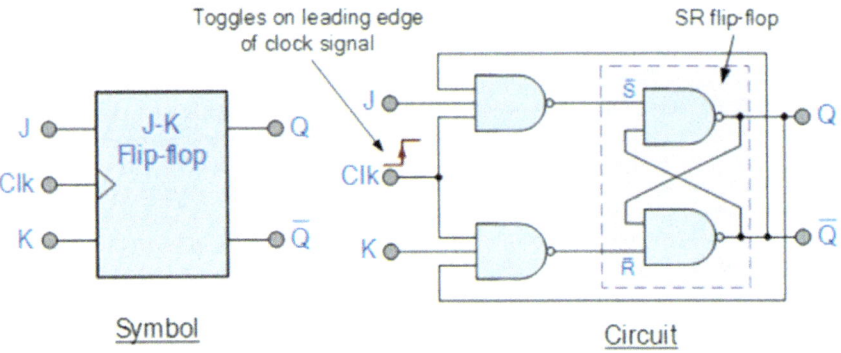

Fig. 3.10 Basic JK Flip-flop (Courtesy: Electronic-tutorials)

Replacing both the S and the R inputs of the previous SR bistable by two inputs called J and K respectively, after its inventor **J**ack **K**ilby. Then this equates to: J = S and K = R (Fig. 3.10).

This simple JK flip-Flop is the most widely used of all the flip-flop designs and is considered to be a universal flip-flop circuit. The sequential operation of the JK flip flop is exactly the same as for the previous SR flip-flop with the same "Set" and "Reset" inputs. The difference this time is that the JK flip flop has no invalid or forbidden input states of the SR Latch even when S and R are both at logic 1.

The two 2-input NAND gates of the gated SR bistable have now been replaced by two 3-input NAND gates with the third input of each gate connected to the outputs at Q and $\overline{Q}$. This cross coupling of the SR flip-flop allows the previously invalid condition of S = 1 and R = 1 state to be used to produce a "toggle action" as the two inputs are now interlocked.

If the circuit is now "SET" the J input is inhibited by the "0" status of $\overline{Q}$ through the lower NAND gate. If the circuit is "RESET" the K input is inhibited by the "0" status of Q through the upper NAND gate. Because Q and $\overline{Q}$ are always different, the inputs can easily be controlled. When both inputs J and K are equal to logic 1, the JK flip-flop toggles as shown in the following truth table (Fig. 3.11).

The JK flip-flop is basically an SR flip-flop with additional feedback which enables only one of its two input terminals, either SET or RESET to be active at any one time thereby eliminating the invalid condition seen previously in the SR flip-flop circuit. Also when both the J and K inputs are at logic 1 at the same time, and the clock input is high, the circuit will "toggle" from its SET state to a RESET state, or vice-versa. This results in the JK flip-flop acting more like a T-type flip-flop when both terminals are high. Adding two extra inputs ($\overline{S}$ and $\overline{R}$) to the two SR flip-flops gives an SR-JK flip-flop.

3.3 The JK Flip Flop

Fig. 3.11 The truth table for the JK function

	Input		Output		Description
	J	K	Q_n	Q_{n+1}	
same as for the SR Latch	0	0	0	0	Memory (no change)
	0	0	1	1	
	0	1	0	0	Reset Q→0
	0	1	1	0	
	1	0	0	1	Set Q→1
	1	0	1	1	
toggle action	1	1	0	1	Toggle
	1	1	1	0	

3.3.1 Dual JK Flip-Flop 74LS73

See Fig. 3.12.

The TTL 74LS73 is a Dual JK flip-flop IC, which contains two individual JK type bistable devices within a single chip enabling single or master–slave toggle flip-flops to be made. Other JK flip-flop IC's include the 74LS107 Dual JK flip-flop with clear, the 74LS109 Dual positive-edge triggered JK flip-flop and the 74LS112 Dual negative-edge triggered flip-flop with both set and clear inputs (SR-JK flip-flop) (Table 3.1).

Although this circuit is an improvement on the clocked SR flip-flop it still suffers from timing problems (called "races") if the output Q changes state before the timing pulse of the clock input has had time to go "OFF". To avoid this, the timing pulse period (T) must be kept as short as possible (high frequency). As this is sometimes not possible with modern TTL IC's the much improved **Master–Slave JK Flip-flop** was developed.

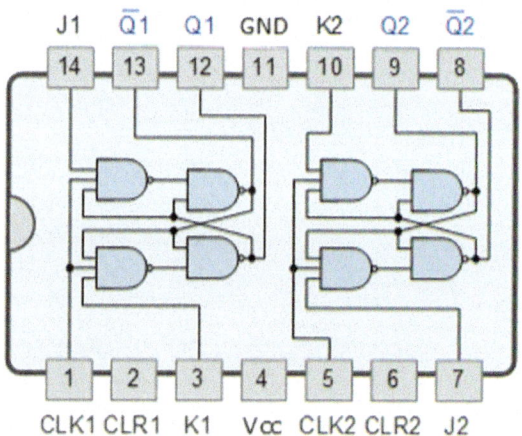

Fig. 3.12 Dual Packages (Courtesy: Electronic-tutorials)

Table 3.1 Other popular JK flip-flop ICs

Device number	Subfamily	Device description
74LS73	LS TTL	Dual JK-type flip flops with clear
74LS76	LS TTL	Dual JK-type flip flops with set and clear
74LS107	LS TTL	Dual JK-type flip flops with clear
4027B	Standard CMOS	Dual JK-type flip flop

3.3.2 The Master–Slave JK Flip-Flop

The **Master–Slave flip-flop** eliminates all the timing problems by using two SR flip-flops connected together in a series configuration with the slave being driven from an inverted clock pulse. One flip-flop acts as the "Master" circuit, which triggers on the leading edge of the clock pulse while the other acts as the "Slave" circuit, which triggers on the falling edge of the clock pulse. This means the device comprises two sections, the master section and the slave section being enabled during opposite half-cycles of the clock signal. The outputs from Q and $\overline{Q}$ from the "Slave" flip-flop are fed back to the inputs of the "Master" with the outputs of the "Master" flip-flop being connected to the two inputs of the "Slave" flip-flop. This feedback configuration from the slave's output to the master's input gives the characteristic toggle of the JK flip-flop as shown in Fig. 3.13.

The input signals J and K are connected to the gated "master" SR flip flop which "locks" the input condition while the clock (Clk) input is high. Because the clock input of the "slave" flip-flop is the inverse (complement) of the "master" clock input, the "slave" SR flip-flop does not toggle. The outputs from the "master" flip-flop are only seen by the gated "slave" flip-flop when the clock input goes low.

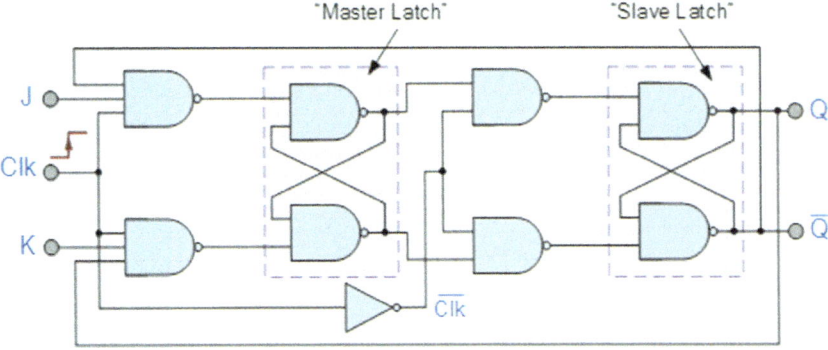

Fig. 3.13 Maste-slave JK Flip-flop (Courtesy: Electronic-tutorials)

3.4 Digital Counters

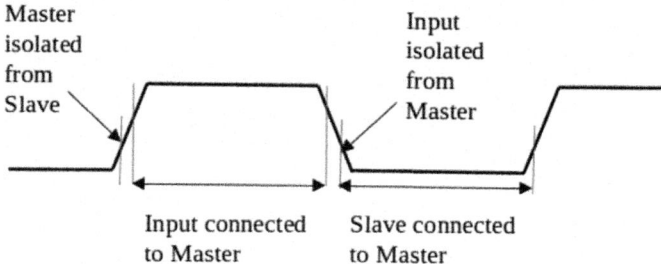

Fig. 3.14 Effects of isolation between stages

When the clock is low, the outputs from the "master" flip-flop are latched and any additional changes to its inputs are ignored. The gated "slave" flip-flop now responds to the state of its inputs passed over by the "master" section (Fig. 3.14).

On the low-to-high transition of the clock pulse, the inputs of the "master" flip-flop are fed through to the gated inputs of the "slave" flip-flop and on the high-to-low transition the same inputs are reflected on the output of the "slave" making this type of flip-flop edge or pulse-triggered.

The circuit accepts input data when the clock signal is high and passes the data to the output on the falling-edge of the clock signal. In other words, the **Master–Slave JK flip-flop** is a "Synchronous" device as it only passes data with the timing of the clock signal.

Normally master slave flip-flops must be configured from two discrete SR flip-flops as with the 7473. Fortunately, there are a number of dual flip-flop chips available, for example: 7467, 7473, 74,107, 74,110, 74,111 etc. and CMOS 4027.

3.4 Digital Counters

Digital counters and shift registers can be designed by Concatenating flip flops. These can run asynchronously or synchronously.

3.4.1 Asynchronous Counters

Asynchronous counters are simple concatenations of flip-flops in toggle mode which are triggered one after the other. That is to say, the Q (or $\overline{Q}$) output on one stage produces the clock pulse for the next stage and so on. Although the generated counter pulses are symmetrical, the initial clock pulse shown in Fig. 3.15 is not. This is not a problem because only the edges of the pulse are important (rising or falling edge). In fact, in the version shown below, the rising edge of the clock switches the first flip-flop. Thereafter, because of the inverted clock inputs, the subsequent flip-flops are toggled by a falling edge.

The outputs are in the form of binary numbers (in the illustrated case 4-bits). Instead of using falling edge triggered flip-flops, rising edge triggered flip-flops can be driven from the $\overline{Q}$ outputs. The only difference between an "up" counter and a "down" counter lies in the chosen outputs. The Q outputs give an up count whereas the down count in given by the $\overline{Q}$ outputs (Fig. 3.16).

Nothing switches from one state to another in zero time. As is clear to imagine, the propagation delay of each stage will accumulate with progression through the counter (Fig. 3.17).

For this reason, a more appropriate design is the synchronous counter.

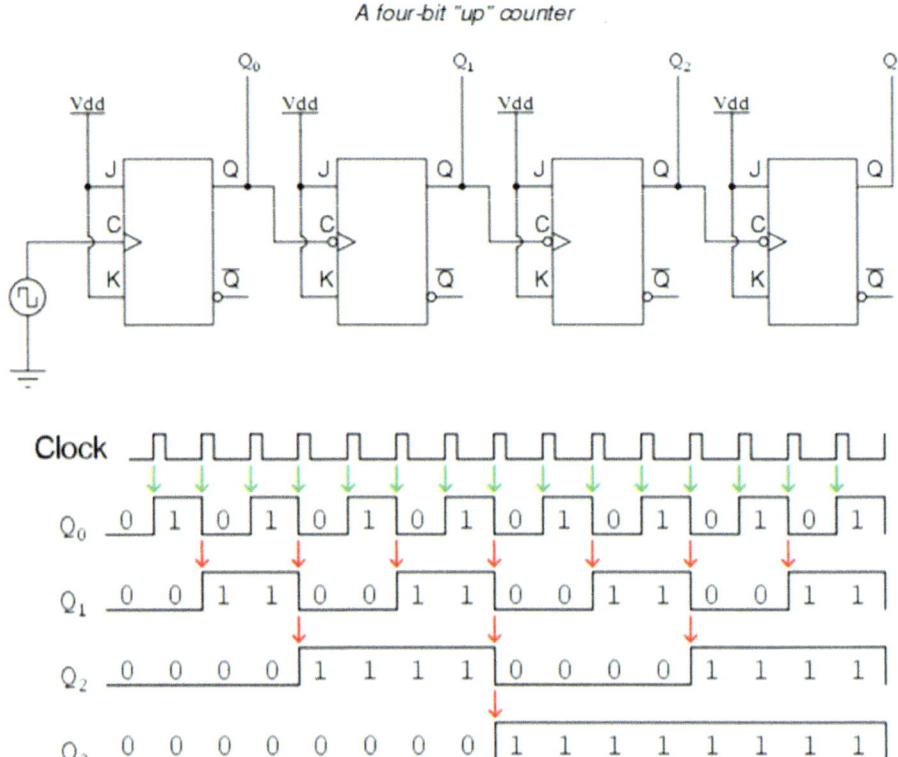

Fig. 3.15 Asynchronous up counter with pulse diagram (Courtesy: All-about-circuits)

3.4 Digital Counters

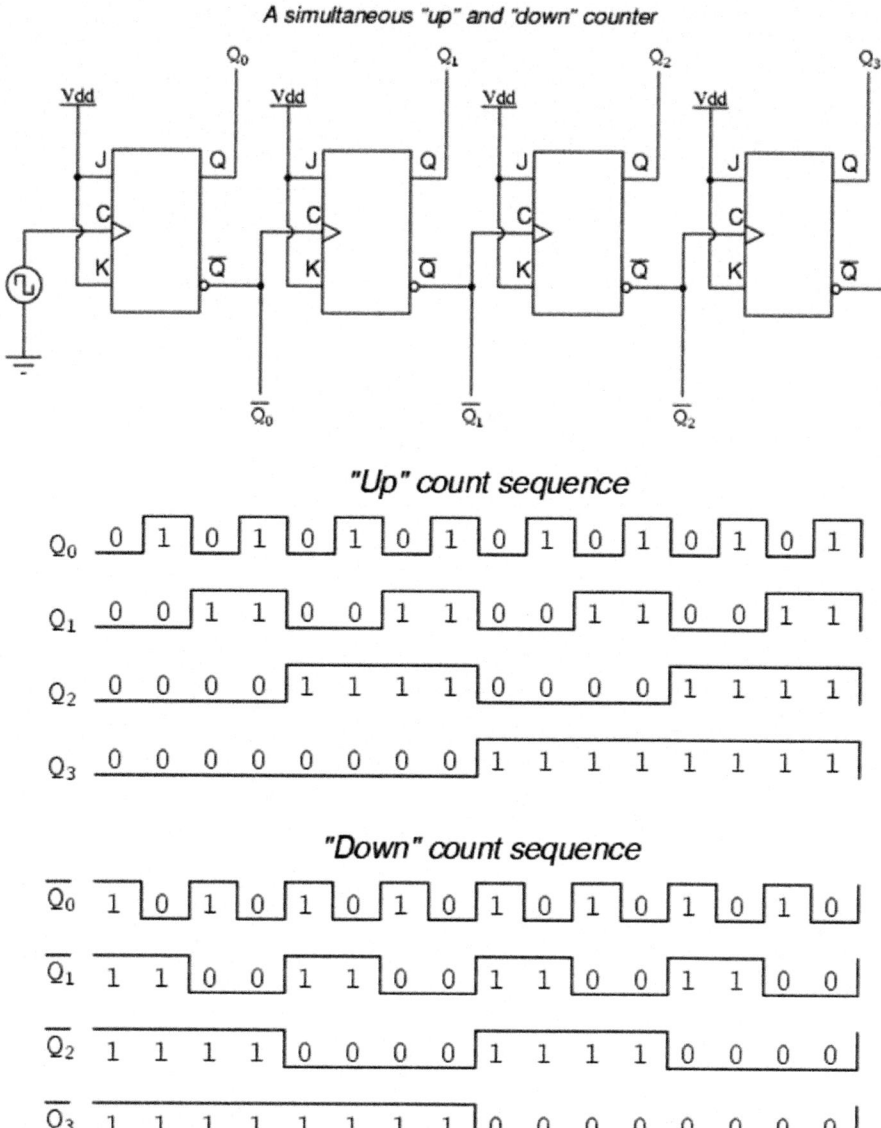

Fig. 3.16 Asynchronous up/down counter with pulse diagram (Courtesy: All-about-circuits)

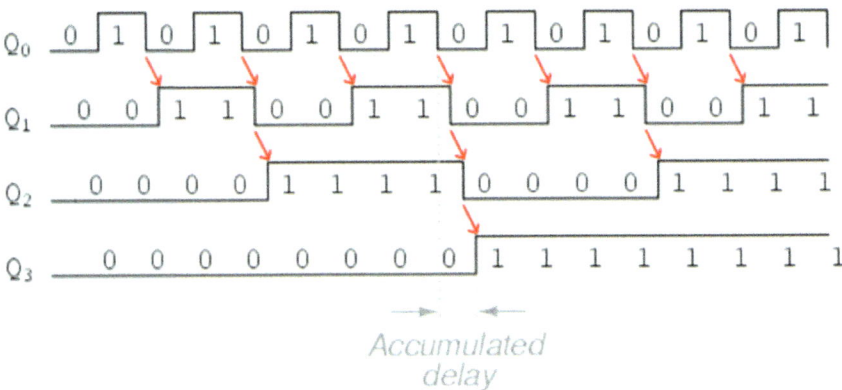

Fig. 3.17 Problems with propagation delay (Courtesy: All-about-circuits)

3.4.2 Synchronous Counters

As shown above, the main problem with asynchronous counters is the propagation delay. This problem is largely overcome by synchronizing flip-flops so that all changes take place simultaneously—not one after the other. However, this is not so straightforward.

One may think that simply adding flip-flops together and driving them all from the same clock simultaneously will result in a synchronous counter. Unfortunately, this is a misconception which will now be demonstrated using a simple 3-bit example of Fig. 3.18.

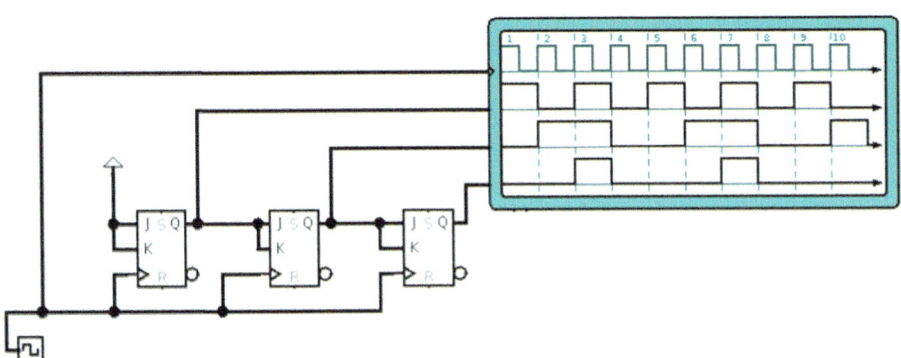

Fig. 3.18 Simple 3-bit synchronous counter

3.4 Digital Counters

Here it can be seen that, although the first two bit trains are correct, the third is wrong. This is due to Qc changing too early because at this point JKc = 1 (Table 3.2)

Clearly some additional logic is required. By making sure that both Qa = 1 and Qb = 1 before Qc can be toggled, the correct sequence is generated as shown in Fig. 3.19.

Of course, this technique can be extended to larger counters and 4-bit examples as demonstrated in Fig. 3.20.

As with asynchronous counters, synchronous counters can count up or down. Its just a question of which outputs to use. In the design below, a little extra logic is employed and an Up/$\overline{\text{Down}}$ input can be used to select the appropriate mode (Fig. 3.21).

Such counters may be extended to almost any number of bits. An example is the 74F161A synchronous 4-bit binary counter.

Table 3.2 Truth table for simple 3-bit synchronous counter

Result			What it should be		
Qa	Qb	Qc	Qa	Qb	Qc
0	0	0	0	0	0
1	0	0	1	0	0
0	1	0	0	1	0
1	1	1	1	1	0
0	0	0	0	0	1
1	0	0	1	0	1
0	1	0	0	1	1
1	1	1	1	1	1
0	0	0	0	0	0

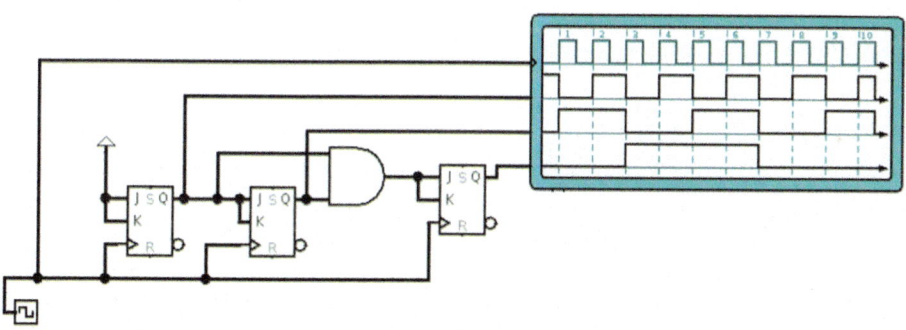

Fig. 3.19 3-bit counter with additional logic

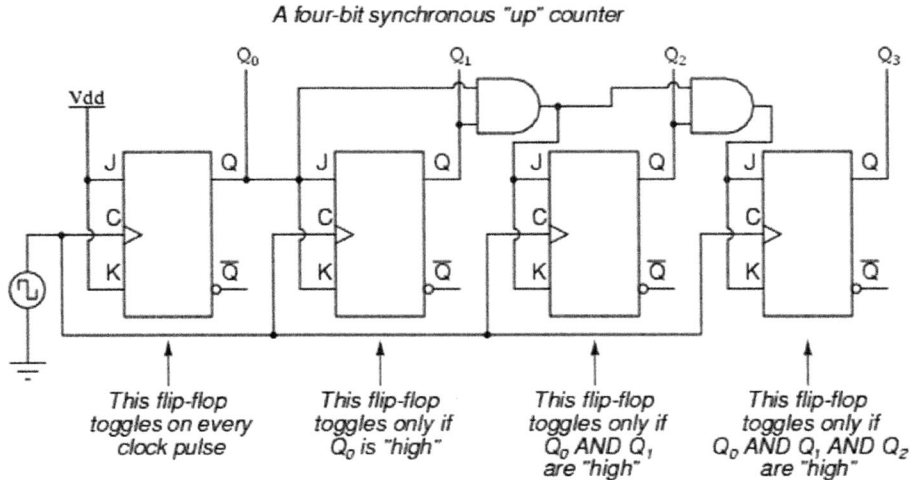

Fig. 3.20 4-bit synchronous up counter [1]

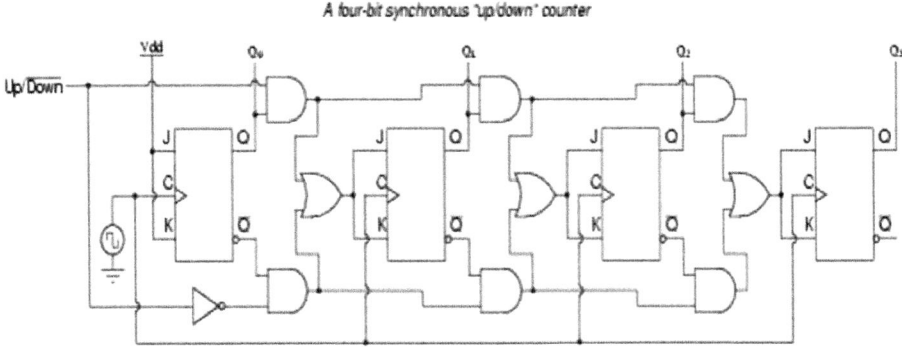

Fig. 3.21 Switchable up/$\overline{\text{down}}$ counter [1]

3.4.2.1 Ring Counters

In addition to up and down counters there are also what are termed ring counters. That is, counters which produce a repeating sequence of semi-symmetric pulses. By connecting a series of D flip-flops together and returning the final Q output to the initial D input, a series of single pulses will be produced. The Set input on the initial flip-flop and the Reset inputs on subsequent flip-flops must be connected to provide an initial digital 1 which will thereafter run in a "ring" around the counter (Fig. 3.22).

3.4 Digital Counters

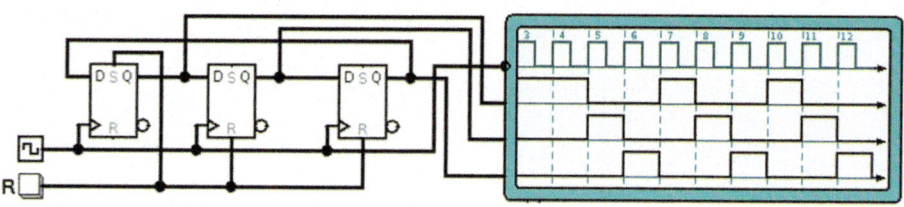

Fig. 3.22 Ring counter using D-Flip-flops

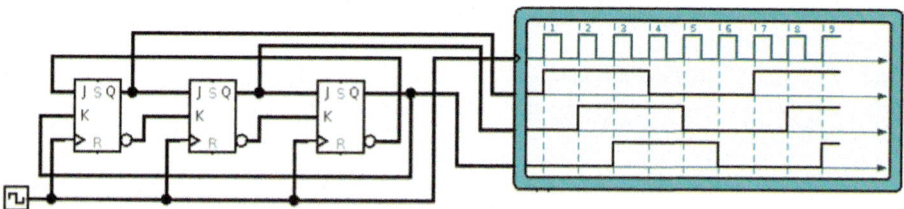

Fig. 3.23 Johnson counter using JK-Flipflops

Another common form is the Johnson Counter (what is sometimes know as the Thermometer code counter). Here no initial condition must be set as the Q and $\overline{Q}$ outputs of the final flip-flop are cross-coupled to the J and K inputs of the first flip-flop (Fig. 3.23).

Such counters can be extended to as many bits as required.

3.4.3 Shift Registers

Shift registers, as the name implies, are sets of registers (flip-flops) which allow a predetermined binary state to be shifted in one direction or another. There are a number of modes in which shift registers can be configured. Serial In—Serial Out (SISO), Serial In—Parallel Out (SIPO), Parallel In—Serial Out (PISO) and Parallel In—Parallel Out (PIPO). These are very effective for shifting binary values left (for multiplication) and right (for division) [2]—see Chap. 4 (Fig. 3.24).

Typically, each stage contains a D Flip-Flop. Commonly used 8-bit SIPO shift registers are the 74LS395 and 74LS164 (Fig. 3.25).

Tri-State means the output can be logic 1, logic 0 or simply floating (high resistance)—useful in connecting to shared bus systems.

There are two functions with shift registers: SHIFT and ROLL. Bits can be shifted left or right and those which are shifted beyond the end of the shift register are lost. On the other hand, bits which are rolled are fed back to the input of the shift register and

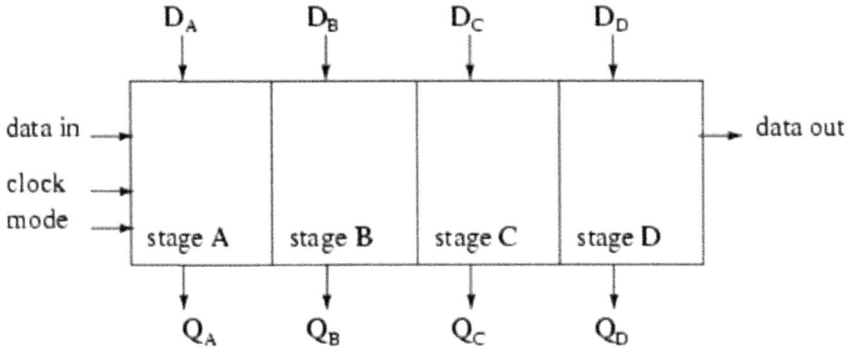

Fig. 3.24 4-bit PIPO Shift register [1]

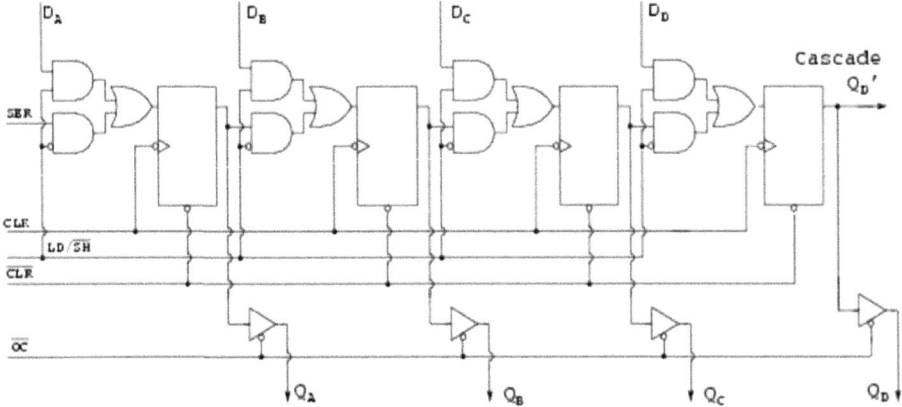

Fig. 3.25 74LS395 PIPO shift register with tri-state outputs (Courtesy: All-about-circuits)

continue. This constitutes a form of serial memory. The ability to roll leads to the concept of the ring counter.

3.4.3.1 Ring Counters

Using feedback, a shift register can be very easily converted into a ring counter (Fig. 3.26).

The parallel D inputs may be used to preselect a starting value (for example 1000). This forms the basis of parallel to serial converters used in serial bus systems (RS232, RS485, Ethernet, etc.).

3.4 Digital Counters

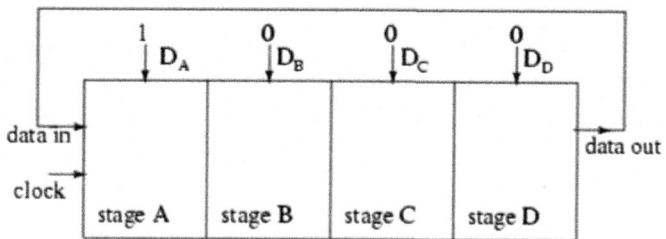

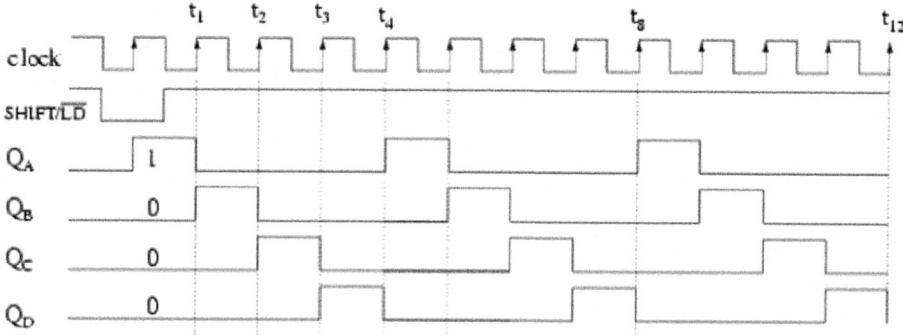

Load 1000 into 4-stage ring counter and shift

Fig. 3.26 Shift register as 4-bit ring counter [1]

3.4.3.2 Johnson Counter

Feeding the inverted output back to the input creates a Johnson counter (Fig. 3.27).

3.4.3.3 Clock Pulse Generation

Whether combinational or sequential logic, time delays play an important role. Often known as propagation delay, each device has a certain time it needs to change state. This problem can also be used to good effect. An oscillator is nothing more than an amplifier with feedback having a phase difference (time delay) between input and output signals.

As the more complicated flip-flops and latches are driven by clock pulses, a means to generate them is necessary. Clocks can easily be built from a number of inverters (NOT gates) (Fig. 3.28).

Quartz crystals are often used as resonant circuits because they are very stable. A simple RC circuit could be used, but the resonance frequency of these tends to drift, particularly with changes in temperature.

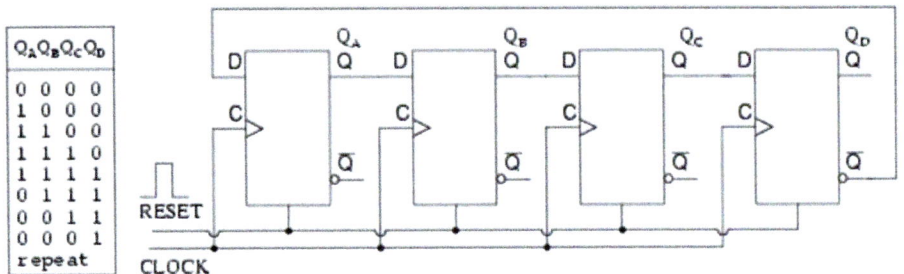

Johnson counter (note the $\overline{Q}_D$ to D_A feedback connection)

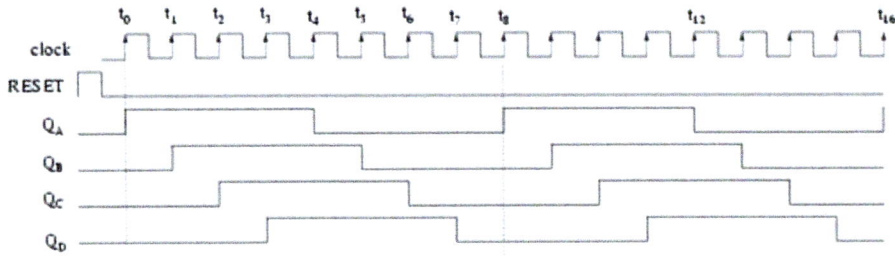

Four stage Johnson counter waveforms

Fig. 3.27 4-bit Johnson counter (Courtesy: All-about-circuits)

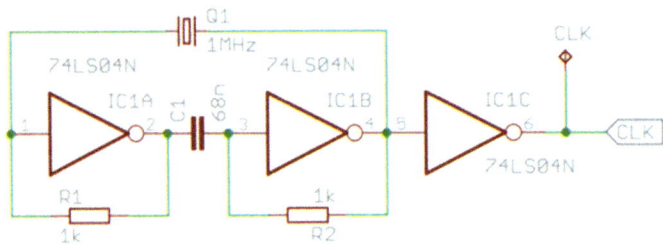

Fig. 3.28 Clock generator using a single 74LS04 chip (Courtesy: Dirk Grappendorf)

3.5 State Diagrams

As shown in Chap. 2, it is possible to analyse digital circuits using a number of techniques including: Boolean algebra, truth tables, Karnaugh maps, circuit and *state diagrams*. State diagrams are particularly good at representing the operation of sequential circuits like counters and shift registers.

A state diagram is a form of digraph where states are represented by circles and transitions between states by arrows. A binary number called the state code is written in the state-circle to indicate the state the machine is in. Directed arrows leaving one state and arriving at another show permissible state transitions. Input variable requirements for transitions are shown immediately next to each transition; the indicated transition will only take place if the input conditions shown are met. Transitions (branches) occur at every clock edge, where the present state is exited and the next-state entered. Given certain input conditions, it is often necessary for the machine to remains in a particular state. This stationary condition is shown as a directed arrow leaving and re-entering the same state. In the partial state diagram shown, the state register contains three flip-flops: if the state register contains '000', then it will remain in that state if A is '0' at the next clock edge; otherwise, if A is '1', it will leave that state.

Figure 3.29 shows the state diagram for a door that can only be opened and closed. It has two states: "opened" and "closed". There are two transitions "open" and "close". Not shown are the "remain" states which would normally be depicted by "self loops" as above.

The following example represents the operation of a toggle switch which has exactly the same function as a T flip-flop (Fig. 3.30).

The four main flip-flop circuits hitherto seen are summarized in state diagram form below. The states shown in the nodes are for the flip-flop Q output and those on the sides for T, D, SR and JK respectively (Fig. 3.31).

Using a D FlipFlop and two EXOR gates, the circuit diagram, truth table and state diagram for the circuit are shown Fig. 3.32.

There are two possible outcomes for Q: 0 and 1 (shown in the nodes). Together with what Q_n is before the clock pulse, the inputs a and b (denoted as pairs on the edges) determine the outcome after the clock pulse Q_{n+1}. These states can be identified from the truth table or by observing the operation of the circuit.

Fig. 3.29 Door example (Courtesy: Real Digital)

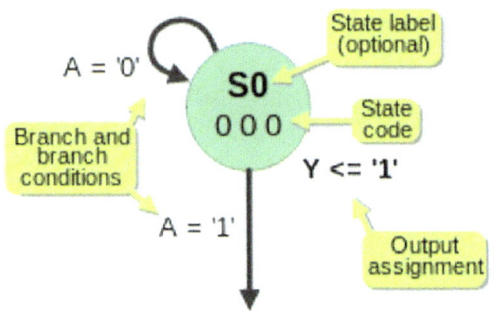

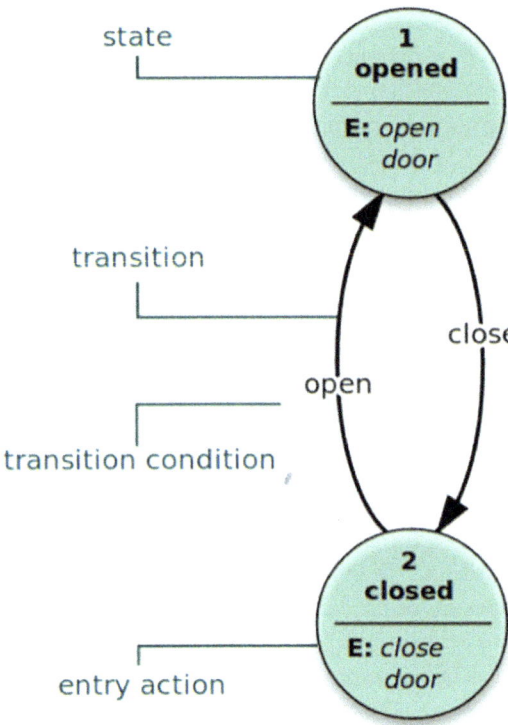

3.5 State Diagrams

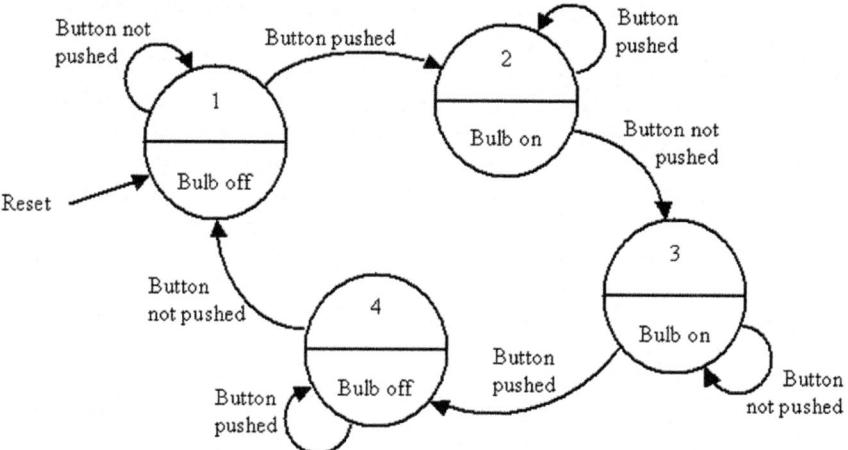

Fig. 3.30 State diagram description of a toggle flip-flop (Courtesy: David Tarnoff)

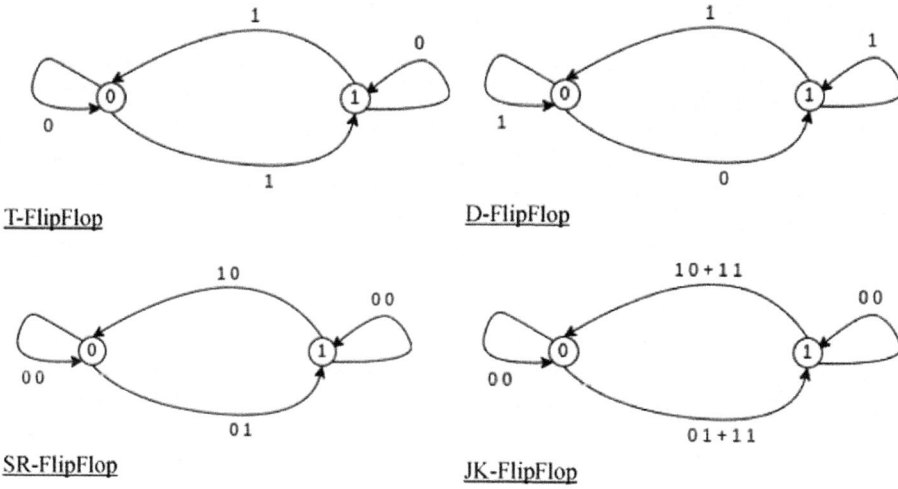

Fig. 3.31 State diagrams for common flip-flops

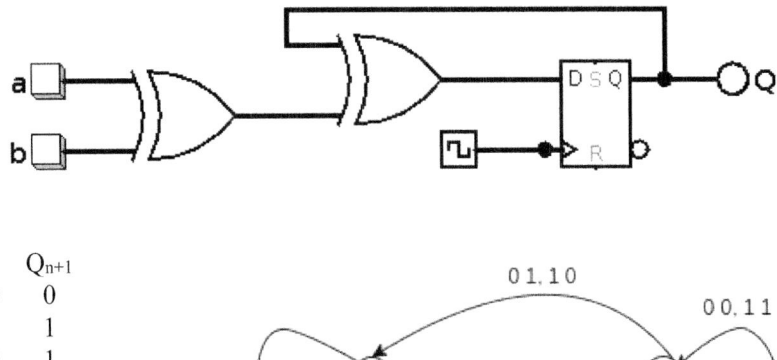

Q_n	a	b	Q_{n+1}
0	0	0	0
0	0	1	1
0	1	0	1
0	1	1	0
1	0	0	1
1	0	1	0
1	1	0	0
1	1	1	1

Fig. 3.32 Example of circuit, truth table and state diagram

Exercises 3—Sequential Logic

Exercise 3.1
Connect the $\overline{Q}$ output with the D input of a D Flip-Flop. Sketch the Q output for the first 4 clock pulses.

Take a look at Exercise3-1.circ.

Exercise 3.2
Draw the circuit diagram using NAND gates and the truth table for a T flip-flop.

3.5 State Diagrams

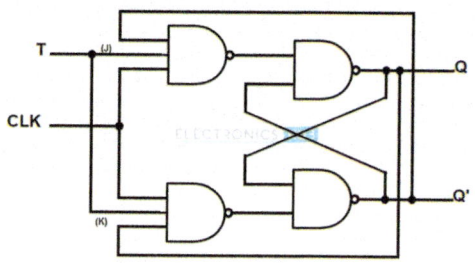

T	Qn	Qn+1	State
0	0	0	No change
0	1	1	No change
1	0	1	Toggle
1	1	0	Toggle

Exercise 3.3

When in the condition S = 0, then only A = 1 and B = 1 or A = 0 and B = 0 gives S = 1.

When in the condition S = 1, then only A = 1 and B = 0 or A = 0 and B = 1 gives S = 0.

Draw the electrical circuit diagram and the corresponding state diagram.

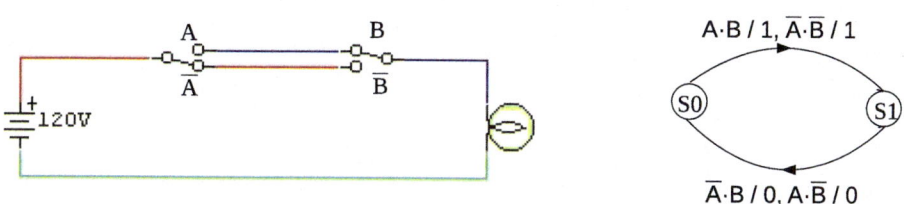

Exercise 3.4

Take a simple 3-bit synchronous counter using JK flip-flops (without additional logic) and work your way step by step through the clock sequence of 8 falling edges. At which point does an error occur?

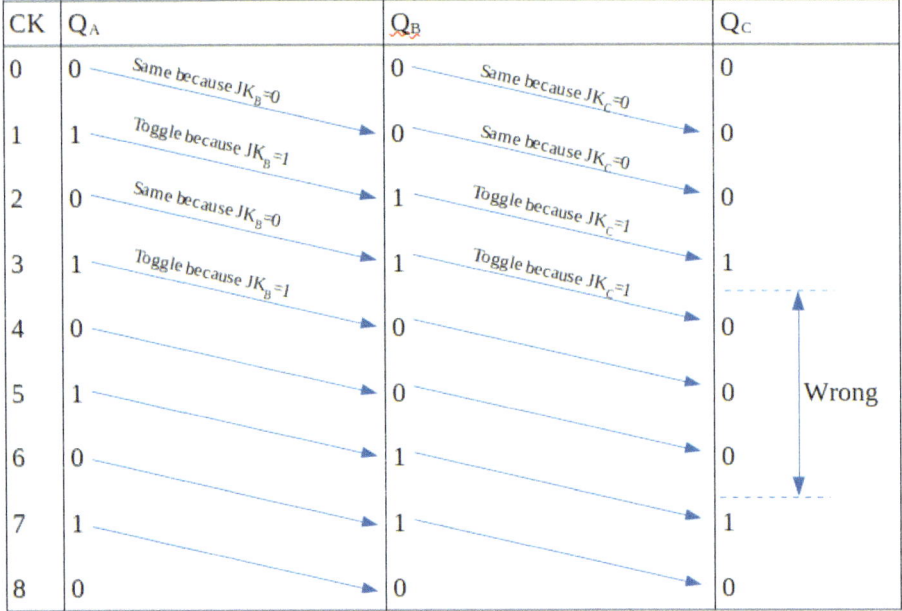

Exercise 3.5
Add additional logic to the 3-bit counter to correct these errors.

....

With additional logic:

3.5 State Diagrams

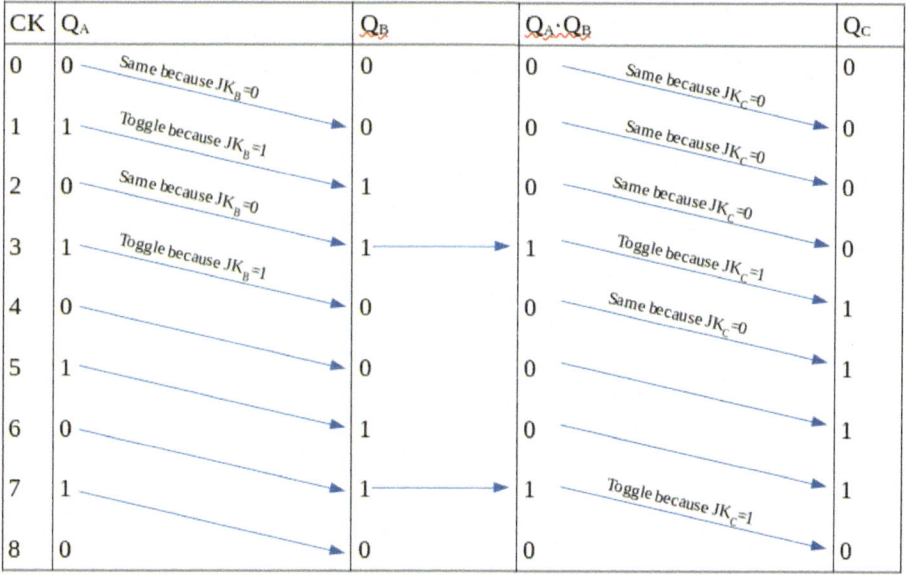

CK	Q_A	Q_B	$Q_A \cdot Q_B$	Q_C
0	0	0	0	0
1	1	0	0	0
2	0	1	0	0
3	1	1	1	0
4	0	0	0	1
5	1	0	0	1
6	0	1	0	1
7	1	1	1	1
8	0	0	0	0

Arrows annotations:
- Q_A column: Same because $JK_B=0$; Toggle because $JK_B=1$; Same because $JK_B=0$; Toggle because $JK_B=1$.
- $Q_A \cdot Q_B$ column: Same because $JK_C=0$; Same because $JK_C=0$; Same because $JK_C=0$; Toggle because $JK_C=1$; Same because $JK_C=0$; Toggle because $JK_C=1$.

Exercise 3.6
Sketch the truth table, Karnaugh map and state diagram for a 4 bit Johnson counter.

....

Truth table   Karnaugh map: State diagram:

D C B A
0 0 0 1
0 0 1 1
0 1 1 1
1 1 1 1
1 1 1 0
1 1 0 0
1 0 0 0
0 0 0 0

CD\AB	00	01	11	10
00	1	0	1	1
01	1	0	0	0
11	1	1	1	0
10	0	0	1	0

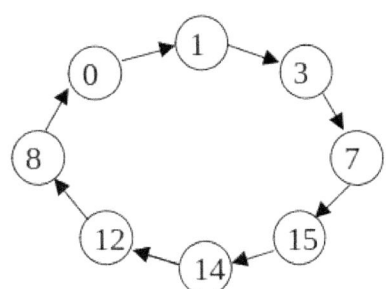

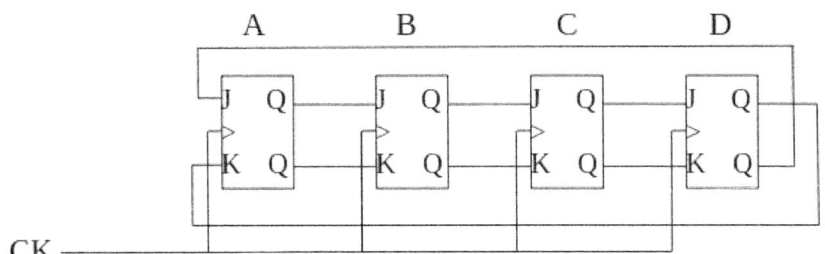

References

1. Kuphaldt. T.R.—Fundamentals of Electrical Engineering and Electronics. Delphi-Kurs, 19 Nov. 2010.
2. Ercegovac. M.D. & T. Lang, Digital Systems and Hardware/Firmware Algorithms, Wiley 1985.
3. Fletcher W.I.—*An engineering approach to dital design*—Prentice-Hall, 1980.
4. Tokheim R.L.—Digital Principles—McGraw-Hill, 1988.

Internet Sites

5. https://www.allaboutcircuits.com/textbook/digital/chpt-12/
6. http://www.cburch.com/logisim/de/download.html
7. http://users.cs.fiu.edu/%7Eprabakar/cda4101/Common/notes/lecture09.html
8. http://www.daenotes.com/flip-flop
9. https://www.electronics-tutorials.ws/sequential/
10. https://faculty.etsu.edu/tarnoff/ntes2150/statemac/statemac.htm
11. https://www.grappendorf.net/projects/6502-home-computer/clock-generation.html
12. https://www.realdigital.org/doc/fc26cf6e35d2a61c6e2871dd9be9e21a

Binary Mathematics

Numbers can be represented in any base. The decimal system is usual because humans have 10 fingers. Computers can only deal with binary because the simplest electrical signals are off and on (2 states). Octal and hexadecimal can easily be changed into binary and are more versatile for people to work with than binary.

Change of Base

The Horner system [Horner, 1819] uses a general algorithm which allows conversion between bases:

$$z_n B^n + z_{n-1} B^{n-1} + \cdots + z_0 B^0 = \sum_{i=0}^{n} z_i B^i$$

Example

Convert binary ($B = 2$) 1010_2 to decimal.

$$z_3 B^3 + z_2 B^2 + z_1 B + z_0$$

$$1 \times 2^3 + 0 \times 2^2 + 1 \times 2 + 0 = 8 + 0 + 2 + 0 = 10_{10}.$$

Example

Convert 24_8 in octal ($B = 8$) to decimal and then into binary.

$$z_1 B + z_0$$

$$2 \times 8 + 4 = 20_{10}$$

From octal to binary is simple:

$$24_8 = 2 \text{ in binary } (010_2) \text{ and } 4 \text{ in binary } (100_2) = 010100_2.$$

Convert back to decimal:

$$\begin{array}{cccccc} 32 & 16 & 8 & 4 & 2 & 1 \\ 0 + & 1 + & 0 + & 1 + & 0 + & 0 \end{array} = 20_{10}$$

Computationally

In Octave [1] or MatLab [Lockhart. S. & E. Tilleson, 2017]: N = dec2base(D,base).
For example: >>N = dec2base(1234,2).

$$N = 10011010010$$

Most other computer programming languages have something similar.

Binary

The right most bit of a binary value is known as the *least significant bit* (LSB) and the left most bit the *most significant bit* (MSB). In most digital technology cases, conversion from binary to decimal and decimal to binary are needed. Conversion from decimal to binary can easily be carried out using successive division by 2.

Example
Convert 205_{10} into binary:

205/2	= 102	Remainder 1
102/2	= 51	Remainder 0
51/2	= 25	Remainder 1
25/2	= 12	Remainder 1
12/2	= 6	Remainder 0
6/2	= 3	Remainder 0
3/2	= 1	Remainder 1
1/2	= 0	Remainder 1

Collecting the results from the bottom to the top gives: 11001101_2.

Conversion from binary into decimal is achieved by successive multiplication by 2 (from right to left) and adding.

4 Binary Mathematics

Example
Convert 11001101_2 back into decimal:

$$128 \quad 64 \quad 32 \quad 16 \quad 8 \quad 4 \quad 2 \quad 1$$
$$1 + 1 + 0 + 0 + 1 + 1 + 0 + 1 = 205_{10}$$

Other often used representations are octal and hexadecimal. Octal values can be converted to 3 digit binary representations and hexadecimal with 4 digits.

Octal

Octal is simply base 8 (i.e. the values 0 to 7). Consequently, conversion from decimal to octal is achieved by using successive division by 8.

Example
Convert 173_{10} to octal:

173/8 = 21 Remainder 5
21/8 = 2 Remainder 5
2/8 = 0 Remainder 2

Collecting the results from the bottom to the top gives: 255_8.
Convert back to decimal:

$$64 \quad 8 \quad 1$$
$$2 \quad 5 \quad 5 = 2 \times 64 + 5 \times 8 + 5 \times 1 = 173_{10}$$

Converting octal values into binary is simple.

Example
Octal: 2   5   5.
Binary: 010   101   101 $= 010101101_2$
Convert back to decimal:

$$256 \quad 128 \quad 64 \quad 32 \quad 16 \quad 8 \quad 4 \quad 2 \quad 1$$
$$0 + 1 + 0 + 1 + 0 + 1 + 1 + 0 + 1 = 173_{10}$$

Hexadecimal

Hexadecimal is base 16 which means all the numbers 0 to 9 plus letters A, B, C, D, E and F used to represent the numbers 10, 11, 12, 13, 14 and 15 respectively. Consequently, conversion from decimal to hexadecimal is achieved by using successive division by 16.

Example

Convert 173_{10} into hexadecimal:

$173/16 = 10$ Remainder 13 (= D)
$10/16 = 0$ Remainder 10 (= A)

Collecting the results from the bottom to the top gives: AD_{16}.
Convert back to decimal:

$$\begin{array}{cc} 16 & 1 \\ A & D \end{array} = 10 \times 16 + 13 \times 1 = 173_{10}$$

Converting hexadecimal values into binary is just simple as it is for octal.

$$\begin{array}{ll} \text{Hexadecimal:} & A \quad D \\ \text{Binary:} & 1010 \; 1101 = 10101101_2 \end{array}$$

Convert back to decimal:

$$\begin{array}{cccccccc} 128 & 64 & 32 & 16 & 8 & 4 & 2 & 1 \end{array}$$
$$1 + 0 + 1 + 0 + 1 + 1 + 0 + 1 = 173_{10}$$

Binary Addition

Binary addition is just as it is with decimal but with only 0 and 1.

Example

Binary addition of 1110_2 and 1100_2.

$$\begin{array}{ll} 1110 & = 14_{10} \\ \underline{1100} & \underline{= 12_{10}} \\ = 11010 & = 26_{10} \end{array}$$

Binary Subtraction

Binary subtraction can also be achieved in the same way as in decimal.

Example

$$\begin{array}{rl} 00111001 = & 57 \\ \underline{-00011110 =} & \underline{-30} \\ = 00011011 = & 27 \end{array}$$

However, digital subtractors are not always available in processors, and computers have difficulty with such procedures. Because the add and invert functions are always available,

4 Binary Mathematics

subtraction can be achieved though complement and addition. However, there are two forms: 1's complement and 2's complement.

Subtraction with 1s Complement
1s complement of the negative number -00011110 is 11100001.
   00111001
$+ \ 11100001$
$= 100011010$
           1   end around carry
$= 00011011$

Subtraction with 2s Complement
2s complement of the negative number $-00011110 = 11100001 + 1 = 11100010$
   00,111,001
$+ \ 11100010$
$= 100011011$
              drop any carry
$= 00011011$

One's Complement and Two's Complement
One's complement and two's complement are two important binary concepts. Two's complement is especially important because it allows us to represent signed numbers in binary, and one's complement is often the interim step to finding the two's complement.

Two's complement also provides an easier way to subtract numbers using addition instead of using the longer, more involved longhand subtraction.

One's Complement
The one's complement of a number is formed by inverting all the bits in a byte by changing each 1 to 0 and each 0 to 1.

Original value	One's Complement
10011001	01100110
10000001	01111110
11110000	00001111
11111111	00000000
00000000	11111111

Two's Complement (Binary Additive Inverse)
Two's complement is a method of representing positive and negative integer values in binary. A two's complement representation automatically includes the sign bit. To form the two's complement, add 1 to the one's complement.

Example

Convert 10011001 to two's complement:

The one's complement is: 01100110.

The two's complement is: 01100110 + 1 = 01100111.

It is however possible for the two's complement to be identical to the original value:

Example

Convert 10000000_2 (128_{10}) to two's complement:

The one's complement is: 01111111.

The two's complement is: 01111111 + 1 = 10000000_2 (-0_{10}).

This means that the 8 bit binary representation of 128 in decimal is negative zero in two's complement form. This is the reason why in many high level computing languages integers range from -128 to $+127$.

Two's complement allows the representation of signed negative values in binary, so here is an introductory demonstration on how to convert a negative decimal value to its negative equivalent in binary using two's complement.

Example

Convert 65_{10} to binary. Ignore the sign for now. Use the absolute value. The absolute value of -65_{10} is 65_{10} (= 01000001_2).

Convert 01000001 to one's complement.

01000001 → 10111110.

Convert 10111110 to two's complement by adding 1 to the one's complement.

10111110 + 1 = 10111111 (= -65_{10}).

If 01000001 ($+65_{10}$) is added to 10111111 (-65_{10}) the result should be 00000000 (0_{10}). However, it is

100000000_2 which has 9 bits (not 8). This last bit (MSB) is known as a carry or overflow which may be ignored at the moment.

01000001 + 10111111 = 100000000_2 (= 0_{10}).

Ignore the carry bit for now. What matters is that original number of bits (bits 7 to bit 0) are all 0.

Theoretically there is no such thing as negative zero (–0). Nothing is always nothing and does not have a sign. 0_{10} = 00000000 can be converted to two's complement 11111111 + 1 = 1 00000000_2 which still gives 0 in binary for the eight relevant bits. Careful: many high level programming languages (such as C) do differentiate between 0 and −0.

It must be known a priori whether the value is a signed integer or not.

4 Binary Mathematics

Example

$1_{10} = 00000001_2$ in two's complement is $11111110 + 1 = 11111111_2$.

If seen as a simple (non-signed) integer, $11111111_2 = 255_{10}$, not -1_{10}. As a rule, assume that a binary value, such as 11111111_2, is a positive integer unless context specifies otherwise. However, in the case of signed binary values then 11111111_2 is -1_{10}, not 255_{10}.

A Simple Trick!

Simple way to convert to 2s complement:
    Copy directly all numbers from the LSB up to and including the first 1. Then invert the rest.

Example

$$010|1\ 001|100\ 0110|1$$
$$101|1\ 110|100\ 1001|1$$

2s complement provides a signed binary representation so why is 1s complement needed?

Because at the hardware level, the ALU usually works with 1s complement. However, in high level programming languages such as FORTRAN, Algol, C etc. [3], it usually appears as 2s complement.

Binary Multiplication

Binary values double with each shift left (SL) and are halved with each shift right (SR). This makes multiplication and division of binary numbers possible.

From the Horner system we know with $z = x \times y$:

$$z = x \sum_{i=1}^{n} y_i z^i$$

Computationally

Using the following algorithm:
    $z \leftarrow 0$
    FOR i = 0 TO n
        IF $y_i = 1$ THEN $z \leftarrow z + SL(x, i)$
    END

Example
Using the above algorithm, multiply 1010_2 by 101_2.

$z = 0$
$i = 0$ $y_i = 1$ $z = 0 + 1010 = 1010$
$i = 1$ $y_i = 0$
$i = 2$ $y_i = 1$ $z = 1010 + 101000 = 110010$

This essentially means with shift left and addition we can achieve binary multiplication.

Example
Multiply 24 by 5.

$24_{10} \times 5_{10} = 11000_2 \times 101_2$

Starting with the LSB and successively moving step by step to the MSB.

       11000   shiftleft by 0 and multiply by 1 = 11000
       110000   shiftleft by 1 and multiply by 0 = 000000
       1100000 shiftleft by 2 and multiply by 1 = 1100000
                add together $= 1111000 = 120_{10}$

Binary Division

Employing the same scheme as with multiplication but with shift right and subtract, binary division can be achieved.

Computationally
Using the following algorithm:

     R ← x
     FOR i = n TO 0
            IF R - SL(x, i) ≥ 0 THEN y_{n-1} ← 1 and (R ← R − SL(x, i)
            ELSE y_{n-1} ← 0
     END

Example
Divide 110010_2 by 101_2.

Using a combination of shift right, shift left and subtraction:

            i = 3   110010−10100 = 1010   ≥ 0 so $y_0 = 1$
            i = 2   1010−10100   < 0 so $y_1 = 0$
            i = 1   1010−1010 = 0   ≥ 0 so $y_2 = 1$
            i = 0   0−101   < 0 so $y_3 = 0$

Collecting the results from the top to the bottom gives: y = 1010_2 with remainder 0 (= 10_{10} exactly).

Example
Convert to binary and divide 53 by 5:
110101 ÷ 101.
The divisor has 3 bits so the first step is to shift left 3 places and subtract:

$$
\begin{array}{lll}
\text{SL3 } 110101-101000 = 001101 & \geq 0 \text{ so } 1 \\
\text{SR1 } 001101-10100 & < 0 \text{ so } 0 \\
\text{SR1 } 001101-1010 = 0011 & \geq 0 \text{ so } 1 \\
\text{SR1 } 0011-101 & < 0 \text{ so } 0
\end{array}
$$

Collecting the results from the top to the bottom gives: 1010_2 with remainder 11_2 (or in decimal 10_{10} with remainder 3_{10}).

We have seen in Chap. 3 how shift registers allow such SL and SR operations to be performed automatically within a computer processor.

Binary Fractions
Binary fractions are dealt with in the same way as with decimal fractions. However, just as decimal values are 1/10, 1/100, 1/1000, etc. at each step to the right of the decimal point, binary values are 1/2, 1/4, 1/8, etc. Just as binary values double from right to left before a decimal point, following the decimal point they are halved from left to right.

Example
Convert 0.1011_2 to decimal.

$$
\begin{array}{cccc}
0. & 1 0 1 1 \\
0. & 1 \times 0.5 + 0 \times 0.25 + 1 \times 0.125 + 1 \times 0.0625 = 0.6875_{10}
\end{array}
$$

Convert back to binary:

$$
\begin{array}{ll}
0.6875-0.5 & = 0.1875 \text{ result so } 1 \\
0.1875-0.25 & = \text{not possible so } 0 \\
0.1875-0.125 & = 0.0625 \text{ result so } 1 \\
0.0625-0.0625 & = 0 \text{ result so } 1
\end{array}
$$

Collecting the results from the top to the bottom gives: 1011, i.e. 0.1011_2.

Data Storage

In a world of 0s and 1s there are no decimal points. Consequently, in real computers values are stored as mantissa and exponent so no explicit decimal point is needed.

In computer jargon, the usual mathematical expressions such as 2×10^6 would be expressed as 2E6 and 4.5×10^{-7} as 45E-8. That way we simply have two integers to store.

Depending of the type declaration (CHAR, INT, LONG etc.), values are stored as one or more 8, 16, 32, etc. bit words with separate mantissa and exponent.

Example

Express 10.6_{10} in two 8-bit words.

10.6 can also be expressed as 106×10^{-1} or 106E-01

Converting 106 and –01 to binary gives:

Mantissa   Exponent
01101010   10000001

Note: the 1 at the beginning of the exponent means that it is negative (complement)!

For larger values, such as a 32-bit word of the data type FLOAT, a 24-bit mantissa (1 sign bit and 23 data bits) and 8-bit exponent (1 sign bit and 7 data bits) would normally be employed.

Exercises 4—Binary Mathematics

Exercise 4.1

Under what conditions are the following two equations true?

$10 + 10 = 100$
$10 * 10 = 100$

....

$10_2 + 10_2 = 100_2$
$10_2 * 10_2 = 100_2$

4 Binary Mathematics

Exercise 4.2
Convert decimal 1284 into Binary, Octal and Hex.

1284_{10}

1284/2 = 642 Rest 0
642/2 = 321 Rest 0
321/2 =160 Rest 1
160/2 = 80   Rest 0
80/2   = 40   Rest 0
40/2   = 20   Rest 0
20/2   =10   Rest 0
10/2   = 5   Rest 0
5/2   = 2   Rest 1
2/2   = 1   Rest 0
1/2   = 0   Rest 1

gives 10100000100_2

Check: Given (from LSB to MSB): 1, 2, 4, 8, 16, 32, 54, 128, 256, 512, 1024…..

$4 + 256 + 1024 = 1284_{10}$

In Octal: $010\ 100\ 000\ 100_2$
=   2   4   0   4 $= 2404_8$

In Hex: $0101\ 0000\ 0100_2$
=   5   0   4 $= 504_{16}$

Exercise 4.3
Convert 10011010_2 into Decimal, Octal and Hex.

Gegeben (von LSB bis MSB): 1; 2; 4; 8; 16; 32; 54; 128; 256; 512; 1024;…..

 1 0 0 1   1 0 1 0
128   16   8 + 2 $= 154_{10}$

010 011 010
 2   3   2 $= 232_8$

1001 1010
 9   A $= 9A_{16}$

Exercise 4.4

Change 5001_{16} into binary. Complement and back into Hex.

Sketch the result (you may need the help of a German dictionary here).

5001_{16}

Hex:	5	0	0	1
Binary:	0101	0000	0000	0001
Complement:	1010	1111	1111	1110
Hex:	A	F	F	E

Exercise 4.5

Subtract 57_{10} from 120_{10} using Binary 2's Compliment.

$$
\begin{array}{ll}
120 & = 01111000 \\
57 & = 00111001 \\
\overline{57} & = 11000110 \\
\overline{57}+1 & = 11000111 \\
\end{array}
$$

Add: 01111000
       11000111
      =100111111 = 63_{10}

Overflow   Sign
(ignore)   bit (0 = +)

Exercise 4.6

Convert 0.1011_2 into Decimal.

Convert 0.5625_{10} into Binary.

How would 23.84 in binary form be stored in a computer?

4 Binary Mathematics

0.1011_2 in Decimal:

0.5; 0.25; 0.125; 0.0625;....

```
 1     0      1        1
0.5 + 0 +  0.125 + 0.0625    = 0.6875
```

0.5625_{10} in Binary:

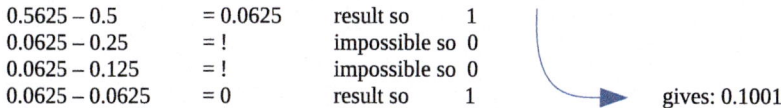

		result so	1
0.5625 − 0.5	= 0.0625		
0.0625 − 0.25	= !	impossible so	0
0.0625 − 0.125	= !	impossible so	0
0.0625 − 0.0625	= 0	result so	1

gives: 0.1001

23,84 would be stored in binary as Mantissa and Exponent separately.

2384×10^{-2}

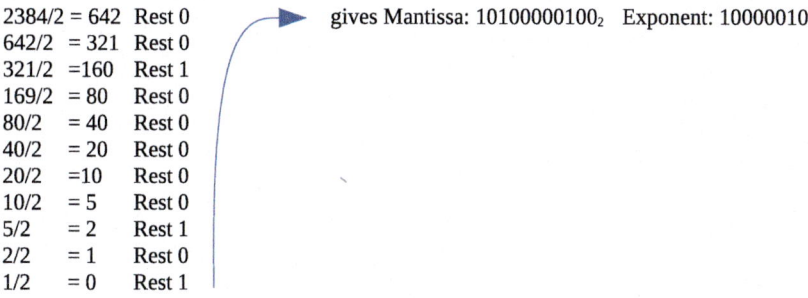

```
2384/2 = 642  Rest 0
642/2  = 321  Rest 0
321/2  =160   Rest 1
169/2  = 80   Rest 0
80/2   = 40   Rest 0
40/2   = 20   Rest 0
20/2   =10    Rest 0
10/2   = 5    Rest 0
5/2    = 2    Rest 1
2/2    = 1    Rest 0
1/2    = 0    Rest 1
```

gives Mantissa: 10100000100_2 Exponent: 10000010

Exercise 4.7

Multiply 42 and 51 in Binary.

Use the appropriate formula to develop an algorithm to multiply 10 by 5 in Binary.

$42_{10} = 101010_2$ $51_{10} = 110011_2$

```
       101010
       110011
       101010
      1010100
    1010100000
   11101000000
   100001011110
```
= 2048 + 64 + 16 + 8 + 4 + 2 = 2142

Using the Horner System: $z = x * \sum_{i=1}^{n} y_i z^i$.

Algorithm:

```
z ← 0
FOR i = 0 TO n
        IF y_i = 1 THEN z ← z + SL(x, i)
END
```

z.B. 1010 * 101   x = 1010 y = 101

```
z ← 0
i ← 0    y_i ← 1    z ← 0 + 1010 = 1010
i ← 1    y_i ← 0
i ← 2    y_i ← 1    z ← 1010 + 101000 = 110010 = 50_{10}
```

Exercise 4.8

Use the binary division method as an algorithm to divide 50 by 5 and repeat for the division of 53 by 5.

```
FOR i = n TO 0
        IF R-SL(x, i) >=0 THEN y_{i-1} ← 1
        AND R ← R – SL(x, i)
        ELSE y_{n-1} ← 0
END
```

50/5

$R \leftarrow 110010$
$x \leftarrow 101$

i=3	110010-101000 = 1010	>= 0	∴ $y_0 = 1$	R = 1010
i=2	1010-10100	< 0	∴ $y_1 = 0$	
i=1	1010-1010	= 0	∴ $y_2 = 1$	R = 0
I=0	0-101	< 0	∴ $y_3 = 0$	

y = 1010   R = 0

53/5

$R \leftarrow 110101$
$x \leftarrow 101$

i=3	110101-101000 = 1101	>= 0	∴ $y_0 = 1$	R = 1010
i=2	01101-10100	< 0	∴ $y_1 = 0$	
i=1	01010-1010 = 0011	>= 0	∴ $y_2 = 1$	R = 0
I=0	0011-101	< 0	∴ $y_3 = 0$	

y = 1010   R = 11

References

1. Hansen. J.S.—*Gnu Octave Beginner's Guide*—PACK Publishers – open source, 30 May 2011.
2. Horner W.G.—A new method of solving numerical equations of all orders, by continuous approximation. *Philosophical Transactions. Royal Society of London.* **109**: 308–335. July 1819. https://doi.org/10.1098/rstl.1819.0023
3. Sebesta, R.W.—*Concepts of programming languages*—10th ed. Addison-Wesley, 2012. {ISBN 978–0–13–139531–2}.
4. Lockhart. S. & E. Tilleson—*An Engineer's Introduction to Programming with MATLAB.*—SDC Publishers, 2017.

The Basic Computer 5

There are many predecessors to the electronic computer, mostly in the form of mechanical calculators. The history is interesting because it shows the obvious demand for such devices whilst demonstrating a number of difficulties which were first overcome with the advent of digital electronics. Mechanical data input was usually complicated and output very limited. In many cases memory systems either didn't exist or were rudimentary punched card or tape formats which restricted any calculation process to a sequence of pre-determined computations.

By connecting the calculator (the CPU) to memory, inputs and outputs via a series of bus systems it became possible to access memory locations randomly rather than in a set order.

5.1 Hardware

A basic electronic computer comprises:

1. A central processing unit (CPU) for control and calculation.
2. A memory system (external to the CPU).
3. An input/output interface.
4. A system of buses to connect the above.

Most modern computers are microprocessor based and use propriety software, though programming at the machine level is not difficult. There are a number of different ways of implementing hardware and executing instructions:

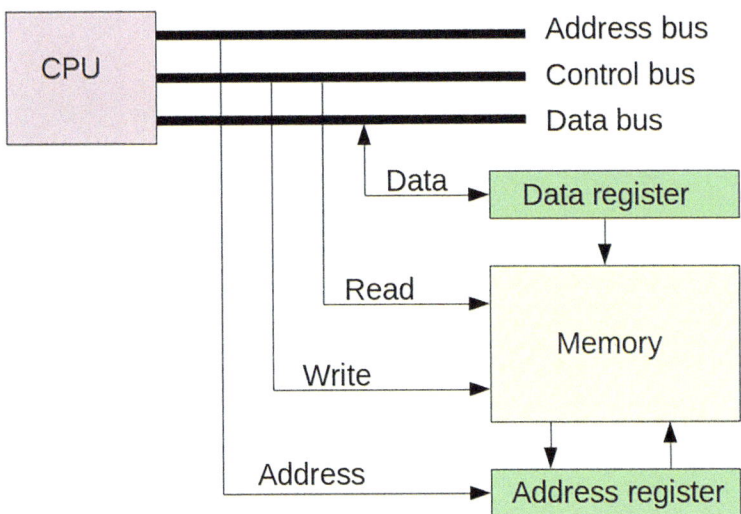

Fig. 5.1 Basic computer architecture

CISC Complex Instruction Set Computer—traditional Von Neumann architecture.
RISC Reduced Instruction Set Computer—faster but greater software overhead.
SISD Single Instruction/ Single Data stream. e.g. Von Neumann Systems (pre 8086).
SIMD Single Instruction/ Multiple Data streams. e.g. Vector multiplication.
MISD Multiple Instruction/ Single Data stream. e.g. Pipe-lining (post 8086).
MIMD Multiple Instruction/ Multiple Data streams. e.g. Full parallel processing.

The Central Processing Unit (CPU) includes an Arithmetic and Logic Unit (ALU) together with a number of working registers and decoders—all connected through buses, as shown in Fig. 5.1.

In Fig. 5.2, the **reset** button forces the processor to start afresh from a known state. The **clock** is the heart beat which keeps everything synchronised. Instructions are given in machine code (via a control bus—not shown) to the **instruction register** (IR) which, via the **instruction decoder**, converts them into micro-instructions. These allow data to be **read** from memory locations given on the **address bus**, into Registers A and B via the **data bus** (using **tri-state** interfaces). The arithmetic and logic unit (ALU) carries out mathematical operations on the data from Registers A and B and places the result in Registers C. These results are **written** to memory given by the **address bus** via the **data bus**. The Test is used to detect certain conditions (compare, overflow, etc.).

The central processor is the core of any computer and the arithmetic and logic unit (ALU) is the core of the processor.

5.2 Arithmetic and Logic Unit (ALU)

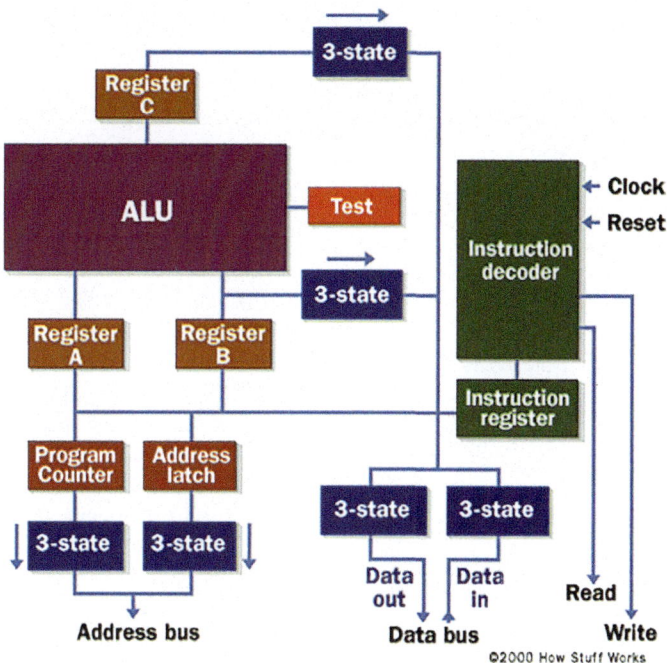

Fig. 5.2 CPU showing interconnections (Courtesy: How stuff works)

5.2 Arithmetic and Logic Unit (ALU)

The ALU comprises a number of full adders connected together to form a 4 (74181), 8, 16, 32 or 64 bit adder, depending on the size of the computer. In addition, there are shift registers (for multiplication and division), the ability to complement (for subtraction) and comparators (Fig. 5.3).

Within the ALU there is an accumulator. This is a small register through which all mathematical operations are carried out. In larger processors there may be several accumulators (registers).

Associated with the ALU are the **instruction register** (IR)—where the current program instruction is stored and the Program Counter (PC)—the memory location where the current program instruction is stored.

Expanding the 1-bit ALU seen earlier to a 2-bit circuit:

Figure 5.4 depicts a simple example arithmetic logic unit (2-bit ALU) that does AND, OR, XOR, and addition.

The CPU follows the following (so called Von Neumann) cycle, repeatedly:

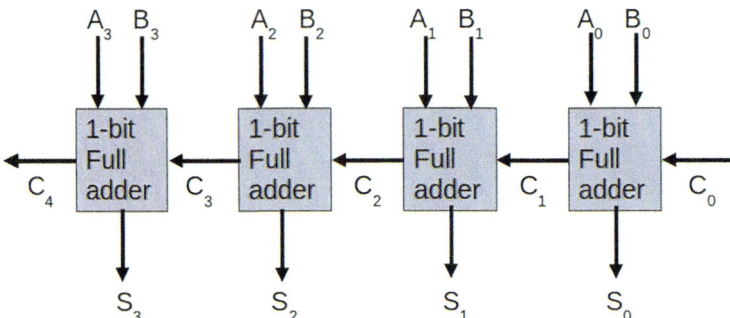

Fig. 5.3 Concatenated 1-bit adders

1. FETCH the instruction from the memory location pointed to by the PC, and copy it to the IR.
2. INTERPRET the instruction in the IR by deciding which circuits for calculation to use.
3. INCREMENT the PC to point to the next memory location.
4. EXECUTE the instruction in the IR, using the circuits selected in the INTERPRET step.

In the case of a "GOTO" or "IF" statement, the value in the PC can be changed, causing the program to go to another memory location.

If the instructions was not a STOP instruction, return to Step 1 and continue.

The IR has two parts:

1. The "Operation Code" or Op Code, a binary number which tells whether to add, subtract, divide, multiply, or some other operation.
2. The memory location of the data to be used.

The Op Codes are broken down further into microprograms or micro-instructions. These micro-instructions contain very basic binary values which tell the processor exactly what to do step by step. They can be implemented in hardware (hard wired) or in software. However, these micro-instructions are usually invisible to the user (Fig. 5.5).

The Von Neumann "fetch–execute" cycle was the means of operation for all processors until the advent of the 8086 microprocessor in 1978.

Intel 8086 Microprocessor

The 8086 [1] was the first microprocessor which deviated from the traditional Von Neumann (fetch–execute) cycle. It comprises a Bus Interface Unit (BIU) and an Execution Unit (EU) which operate in parallel. This is not real parallel processing as in the form

5.2 Arithmetic and Logic Unit (ALU)

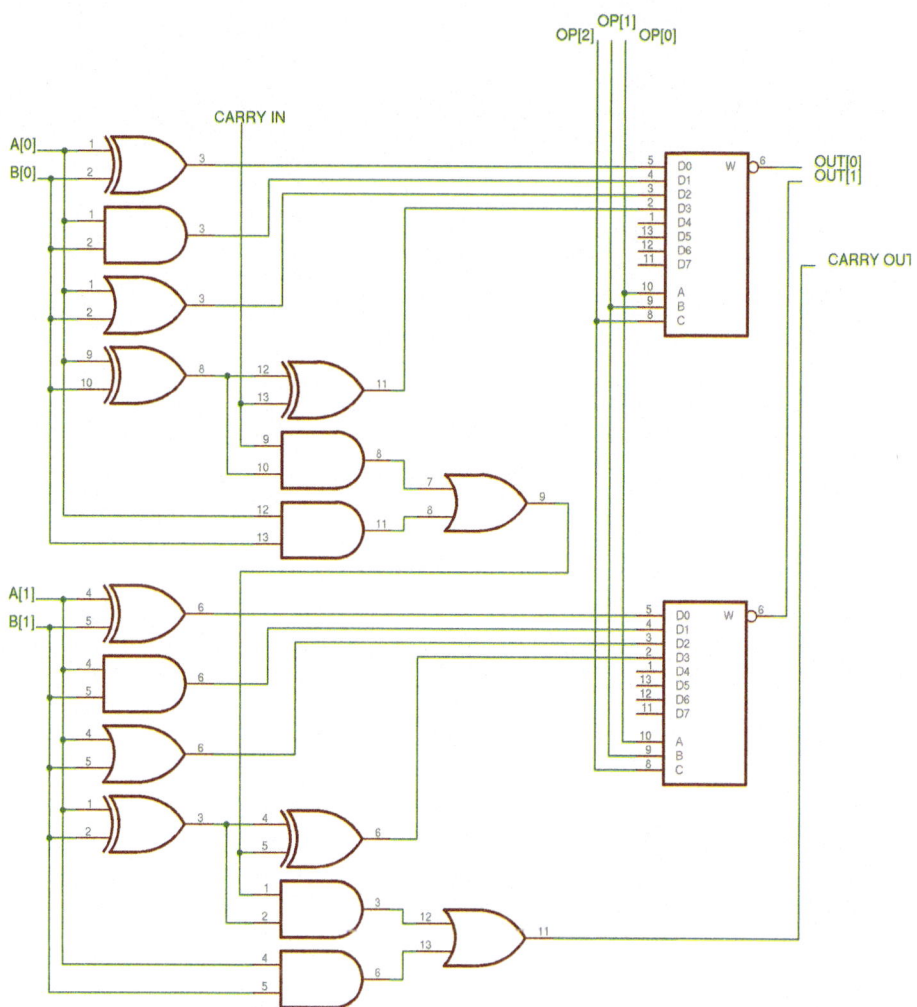

Fig. 5.4 3-bit ALU (Courtesy: commons.wikimedia.org)

of multiprocessor architecture but does allow a degree of parallel operation known as "pipe-lining". It was a 16-bit processor initially launched in 1978 in 5, 8 and 10 MHz versions.

The block diagram of Fig. 5.6 shows how the BIU fetches instructions or data from memory and is able to read and write data to memory or ports. The EU tells the BIU from where to fetch the instructions or data, decodes and executes the instructions.

In order to increase the execution speed, the BIU fetches as many as six instruction bytes ahead of time from memory. All six bytes are then held in a 6 byte first in first out (FIFO) register called the instruction queue. Then all bytes are passed on one by one to

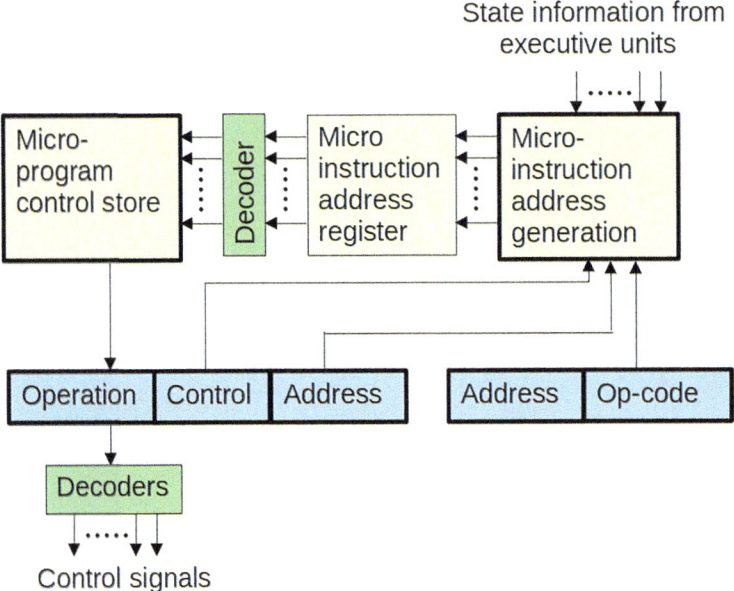

Fig. 5.5 Micro-instructions generated as a result of Op-code execution

the EU. This pre-fetching operation of the BIU may be carried out parallel to execution operations of the EU, which improves the processing speed of the instruction.

The EU contains the necessary control circuitry to perform various internal operations. A decoder in the EU decodes the instructions fetched from memory to generate different internal or external control signals required to perform the operation.

The 8086 has 4 general purpose registers (AH/AL, BH/BL, CH/CL and DH/DL) which can be used as 8-bit registers individually or in pairs as 16-bit registers. Each register contains a High and a Low 8-bit byte.

AH and AL form the basic accumulator register that stores operands for arithmetic operation like divide, rotate, etc.

BH and BL are mainly used as a base register. It holds the starting base location of a memory region within a data segment.

CH and CL are counter registers. They are primarily used as a loop counter and used in shift and rotate operations.

DH and DL are used in multiplication, division, and I/O operations and contain the I/O port address for I/O instructions.

Additional registers called segment registers generate memory addresses when combined with other registers in the microprocessor. In the 8086 microprocessor, memory is divided into 4 segments as follows:

5.2 Arithmetic and Logic Unit (ALU)

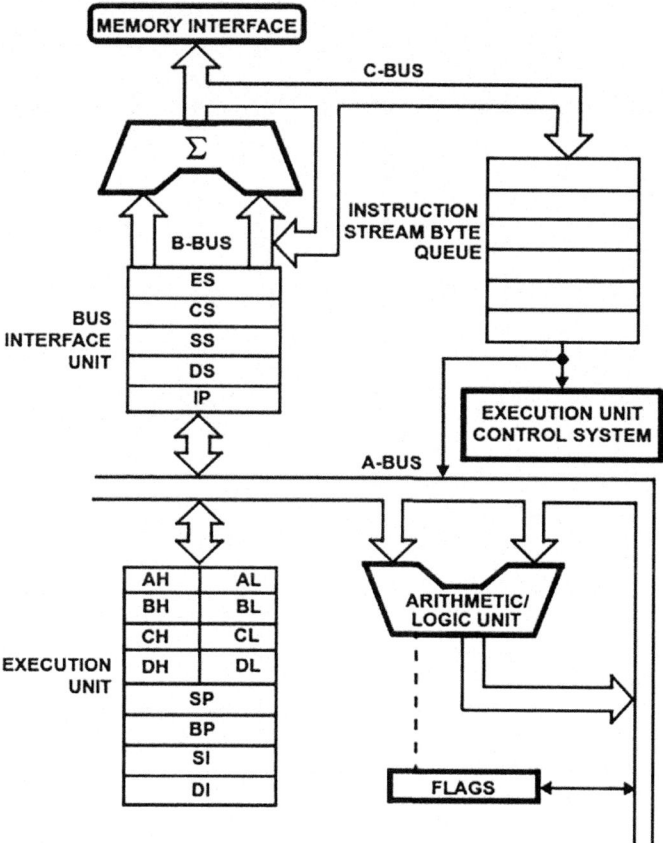

Fig. 5.6 8086 microprocessor

The Code Segment (CS) register is used for addressing a memory location in the Code Segment of the memory, where the executable program is stored.

The Data Segment (DS) register contains most data used by the program. Data are accessed in the Data Segment by an offset address or the content of another register that holds the offset address.

The Stack Segment (SS) register defines the area of memory used for the stack.

The Extra Segment (ES) register is an additional data segment for memory access. When used in conjunction with the BH/BL, SI, or DI registers, the ES register allows for indirect addressing of memory locations within the extra segment. This can be useful for accessing data or code located in different segments of memory.

The Instruction Pointer (IP) register holds the address of the current instruction.

Since the birth of the microprocessor over 60 different devices have been developed and no end is in sight.

Exercises 5–Micro–Computer Systems

Exercise 5.1

Which bus systems are indispensable for every computer?

What does "tri-state" mean and what is it used for?

....

Control bus, data bus, address bus.

Outputs can "float". In high-impedance status (neither 0 nor 1).

Exercise 5.2

What distinguishes the 8086 from the Von Neumann architecture?

....

Von Neumann ("fetch–execute" cycle) is the original processor method. From 8086 onward, "pipelining" is used with a "Bus Interface Unit" (BIU) and an "Execution Unit" (EU). New data is fetched from the memory while the already loaded data is being processed.

Exercise 5.3

Design a computer!

....

Let your imagination run free!

References

1. Rector. R & G. Alexy–The 8086 Book–McGraw-Hill, 1980. {ISBN 0–931988–29–2}.

Internet Sites

2. https://grandidierite.github.io/intel-8086-registers-and-memory-organization-NASM-16-bit/.
3. https://computer.howstuffworks.com/microprocessor.htm.
4. http://www.codeproject.com/Articles/315505/How-processor-assembler-and-programming-languages.
5. https://www.geeksforgeeks.org/computer-organization-hardwired-vs-micro-programmed-control-unit/.

Software 6

At the lowest hardware level, the processor can only deal with binary values (10110000, 01110101, 00000101, 10101100, etc.).

To make this compacter the binary is stored as 8 bit words (bytes) which can be displayed as octal or hexadecimal values (numeric code) as Machine code (B0 75 05 AC F3, etc. usually displayed in Hex)–compare the right-hand column (u... P. t.. etc.) with the ASCII table. For humans to be able to read and program at this level, a set of assembly language mnemonics (LDA 77, STA 63, etc.) are used.

Machine code	Assembly language
0EDF: 0100 B075	MOV AL, 75
0EDF: 0102 05ACF3	ADD AX, F3AC
0EDF: 0105 AA	STOSB
0EDF: 0106 A00AEB	MOV AL, [EB0A]
0EDF: 0109 06	PUSH ES
0EDF: 010A 3CB2	CMP, AL B2
0EDF: 010C 756D	JNZ 017B
0EDF: 010E 6D	DB 6D
0EDF: 010F 13A80150	ADC BP, [BX + SI + 5001]

0EDF: 0100 B0 75 05 AC F3 AA A0 0A-EB 06 3C B2 75 6D 13   u......
0EDF: 0110 A8 01 50 14 74 B1 BE 32–01 8D 8B 1E 34 00 CE 0E   P.t..2

The instruction for the assembly command MOV AL is B0 in hex. 75 is the value to be loaded into the AL register. This is a two byte command starting in the memory position 0100, hence the next command (a 3 byte command) starts at the memory locations 0102 and continues to location 0104. 05 in hex represents the command ADD AX which tells the processor to add the value F3AC to the AX register. Note that the order of the bytes is reversed between machine code and assembler script. This is not always so but is typical for Intel processors. The same applies to certain mnemonics, for example MOV (move) as shown above often appears as LD or LOAD in some processors.

However, for more advanced programming, a high-level language (FORTRAN, Algol, C/C++ etc.) is employed. In some versions of C [1] it is possible to jump in and out of assembler mode using the asm command.

6.1 The Asm Command in C

Example of C-code with the inclusion of an assembly language routine with the asm command.

```
#include <stdio.h>
int LenCount=3*2;              /* number of bytes in field */
int AddrEnd;                   /* address of field termination */
int Count[]={17, 4, 9};        /* sum of numeric field */
int Sum()
{       asm{
    mov ax, 0                  /* set Sum to 0 */
    mov cx, OFFSET Count       /* calculate end address */
    add cx, LenCount
    mov AddrEnd, cx            /* save end address */
    mov si, OFFSET Count       /* address of current count */
@below:                        /* start of loop */
    add ax, [si]               /* count summation */
    add si, 2                  /* address 2 bytes further */
    cmp si, AddrEnd            /* compare current address with end address */
    jb @below                  /* if smaller then repeat */

    }
}
main() { printf("%d\n", Sum()); }
```

This is not usually available in other high level programming languages intended for hardware independent cross-platform operation.

High level languages run directly via an operating system (which may also be written in assembler or C/C++). In fact, C is integrated into Linux (in which the operating system itself is written). In addition, application programs (MatLab, LabView, Python etc.)

6.1 The Asm Command in C

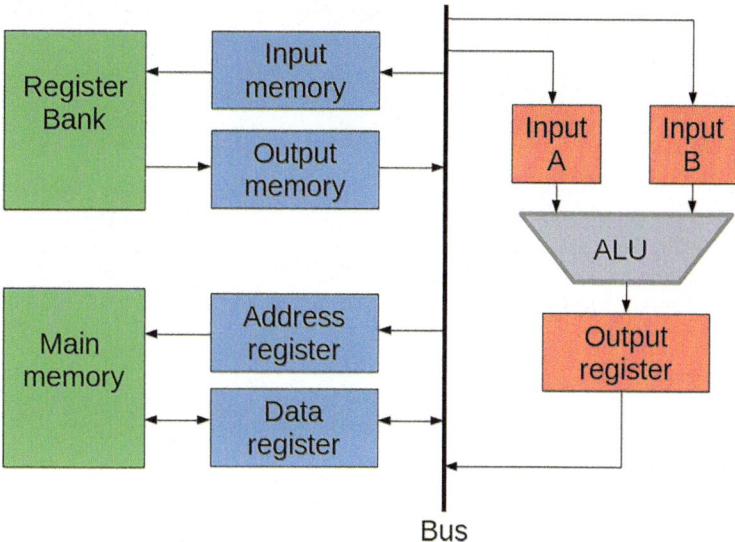

Fig. 6.1 Model of a simple processor

and user software (Libre Office, Word, Excel etc.) are also usually written in high-level languages.

The next section demonstrates a very simple hypothetical processor and its basic function. For simplicity, the three essential bus systems are depicted as one in Fig. 6.1. Each assembler instruction is formed by a concatenation of micro-programs. These are the individual step by step instructions needed to select registers, invoke functions and transfer data.

Example micro-programs available to the microprocessor in hardware form.

RNR   Register select.
RRW   Register bank read/write.
RES   Input register read/write.
RAS   Output register read/write.
MRW   Main memory read/write.
MAR   Address register read/write.
MDR   Data register read/write.
AIN   ALU input register control.
AOU   ALU output register control.
AFK   ALU control function.
BUS   Central bus control.

These microprograms are not available to the user. They (or combinations thereof) are executed by the microprocessor on receipt of an assembler command.

Example

ADD R1, (R2).

1. Read the contents from register R2 in the output register of the register bank.
   RNR, RES

2. Send the contents of the output register over the bus and into the main memory address register.
   RAS, BUS, MRW

3. Bring data from the main memory into the data register.
   MRW, MAR

4. Send the contents of the data register over the bus to input A of the ALU.
   MDR, BUS, AIN

5. Read the contents from register R1 in the output register of the register bank.
   RNR, RAS

6. Send the contents of the output register over the bus to input B of the ALU.
   RAS, BUS, AIN

7. Add the two values held in inputs A and B together, placing the result in the ALU output register.
   AFK, AOU

8. Send the contents of the ALU output register over the bus into the main memory data register.
   AOU, BUS, MDR

…and so on…

There are several potential addressing modes and some of the simpler and commonly available types are described in detail later.

6.2 Basic Simple Assembler

A good example of a very basic assembler code typically found in early microprocessors or modern RISC (Reduced Instruction Set Computer) processors is illustrated in Table 6.1.

Remaining with the simple hypothetical model with a single accumulator, a program written in assembler may look something like that shown in Table 6.2.

6.2 Basic Simple Assembler

Table 6.1 Simple assembly language commands

Numeric code	Mnemonic code	Instruction	Description
1xx	ADD	ADD	Add the value stored in memory location xx to whatever value is currently in the accumulator. Note: the contents of the memory are not changed, and the actions of the accumulator are not defined for add instructions that cause sums larger than 3 digits. Similarly to SUBTRACT, a negative flag could be set in the event of an overflow
2xx	SUB	SUBTRACT	Subtract the value stored in memory location xx from whatever value is currently in the accumulator. Note: the contents of the memory location are not changed, and the actions of the accumulator are not defined for subtract instructions that cause negative results—however, a negative flag will be set so that **7xx (BRZ)** and **8xx (BRP)** can be used properly
3xx	STA	STORE	Store the contents of the accumulator in memory location xx (destructive). Note: the contents of the accumulator are not changed (non-destructive), but the contents held in the memory location are replaced regardless of what was in there previously (destructive)
5xx	LDA	LOAD	Load the value from memory location xx (non-destructive) and enter it in the accumulator (destructive)
6xx	BRA	BRANCH (unconditional)	Set the program counter to the given address (value xx). That is, value xx will be the next instruction executed

(continued)

Table 6.1 (continued)

Numeric code	Mnemonic code	Instruction	Description
7xx	BRZ	BRANCH IF ZERO (conditional)	If the accumulator contains the value 000, set the program counter to the value xx. Otherwise, do nothing. Whether the negative flag is taken into account or not, is undefined. When a SUBTRACT underflows the accumulator, this flag is set, after which the accumulator is undefined, potentially zero, causing the behaviour of BRZ to be undefined on underflow. Likely behaviour would be to branch if accumulator is zero and negative flag is not set. Note: since the program is stored in memory, data and program instructions all have the same address/location format
8xx	BRP	BRANCH IF POSITIVE (conditional)	If the accumulator is 0 or positive, set the program counter to the value xx. Otherwise do nothing. As memory cells can only hold values between 0 and 999, this instruction depends solely on the negative flag set by an underflow on SUBTRACT and potentially on an overflow on ADD (undefined). Note: since the program is stored in memory, data and program instructions all have the same address/location format
901	INP	INPUT	Go to the INBOX, fetch the value from the user and put it in the accumulator. Note: this will overwrite whatever value was in the accumulator (destructive)
902	OUT	OUTPUT	Copy the value from the accumulator to the OUTBOX. Note: the contents of the accumulator are not changed (non-destructive)
000	HLT	HALT	Stop working
	DAT	DATA	This is an assembler instruction which simply loads the value into the next available memory location. DAT can also be used in conjunction with labels to declare variables. For example, DAT 984 will store the value 984 into a memory location at the address of the DAT instruction

6.2 Basic Simple Assembler

Table 6.2 Example assembly language program

Addr	Contents		
1	LDA	200	Load the accumulator with the contents of address 200
2	STA	101	Store the contents of the accumulator in address 101
3	LDA	201	Load the accumulator with the contents of address 201
4	SUB	101	Subtract the contents of address 101 from the accumulator
5	BRN	13	If accumulator negative then branch (jump) to program line 13
6	LDAI	101	Load the accumulator indirectly from address 101
7	ADD	102	Add to the accumulator the contents of address 102
8	STA	102	Store the contents of the accumulator in address 102
9	LDA	101	Load the accumulator with the contents of address 101
10	ADD	100	Add to the accumulator the contents of address 100 (increment)
11	STA	101	Store the contents of the accumulator in address 101
12	BRA	3	Unconditional branch to program line 3
13	HLT		Stop
Addr	Contents		
100	1		First constant
101	~~202~~ 203		Address of current sum
102	0		Temporary/final result
Addr	Contents		
200	202		Address of first sum
201	204		Address of last sum
202	10		First sum
203	7		Second sum
204	4		Last sum

What follows is an example of an assembly language program using direct and indirect addressing. The program comprises 13 instructions in the first 13 memory locations. The memory locations 100 to 102 and 200 to 204 are used for temporary data storage. It is often convenient to precede direct numerical values with a hash (#) to distinguish them from memory locations. The sequence of events is easy to follow:

LDA 200 loads the value held in memory location 200 (not the value 200 itself!) which puts #202, into the accumulator.

STA 101 causes this value (#202) then to be stored in memory location 101. Whatever was there previously will simply be overwritten.

LDA 201 a new value (#204) is then loaded from location 201 into the accumulator.

SUB 101 The contents (#202) of the address 101 are then subtracted from the contents held in the accumulator.

BRN 13 means "branch" or "jump" to location 13 if the value of the accumulator is negative. In this case it is positive, so the program does not branch but continues to step 6.

LDAI 101 uses indirect addressing. This means that the value held in the memory location 101 is used to access the location (202) where the required data (#10) is held. This value (#10) is then passed to the accumulator.

ADD 102 adds the value (#0) held in location 102 to whatever is held in the accumulator (#10).

STA 102 now stores this value (#10) into memory location 102.

LDA 101 loads the value (#202) held in memory location 101 into the accumulator.

ADD 100 adds the value (#1) held in location 100 to whatever is held in the accumulator (#202). Now the accumulator contains the value #203.

STA 101 stores the value in the accumulator (#203) into memory location 101.

BRA 3 is an unconditional branch to program location 3 where the program continues with LDA 201, etc.

HLT: at some point the accumulator will be negative and the branch function at program step 5 will cause the program to jump to location 13 where the program stops.

For those with enough patience (or curiosity) to go through another three loops of the program routine, it will be clear that this is a simple incrementation algorithm. In a high-level language, this would follow a procedure like x = x + 1 or in C simply x++. Branching could be achieved using the usual IF...THEN...ELSE structures or similar.

Normally similar structures are common in high-level languages. In such cases the programs can either be interpreted (each line of code individually converted to machine code and executed) or compiled (the complete program is converted to an executable machine code file).

Addressing Modes

Most processors possess a number of addressing modes. A few of the most common ones are described below.

Immediate
ADDA, #8 $A \leftarrow A + \#8$

Accumulator takes the value 8.

6.2 Basic Simple Assembler

Note: Immediate addressing is not always available.

Direct

ADD A, 1234 $A \leftarrow A + (1234)$

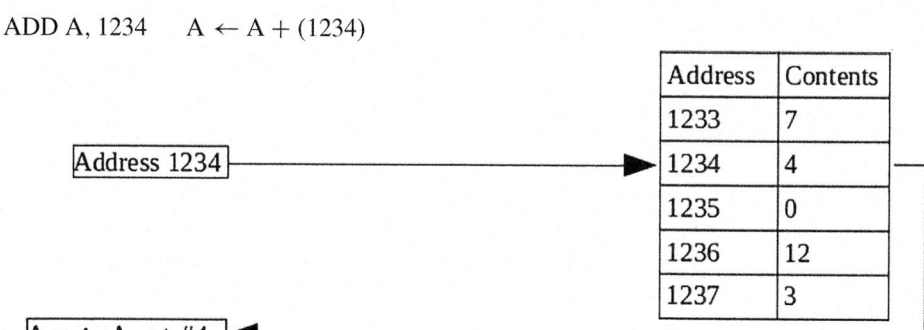

The value held in the location addressed is added to the Accumulator.

Indirect
ADDi A, 1234 $A \leftarrow A + ((1234))$

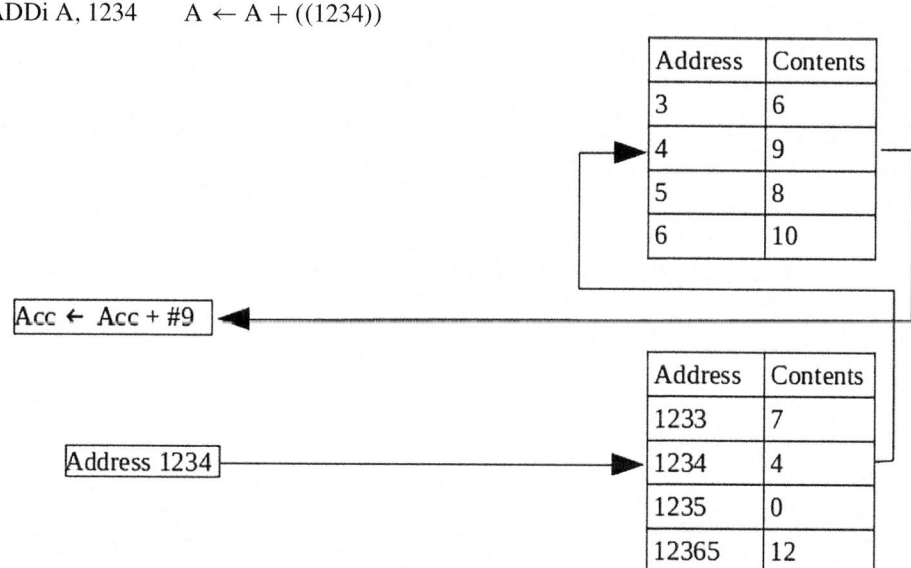

Added to the Accumulator is the value held in the location addressed by the contents of the initial addressed location.

Relative
ADDr A, PC # 1234 $A \leftarrow A + (PC + 1234)$

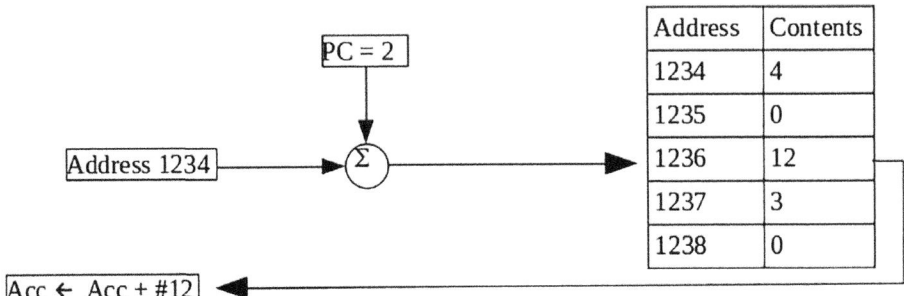

Added to the Accumulator is the value held in the location addressed by the address plus the contents of the program counter (PC).

Indexed
As for Relative but instead of the program counter (PC) the index register (IR) is used.

Note the similarity to using pointers in C!

For the 8086 there are far more addressing modes.

6.3 Compilers

A Compiler comprises a Lexer and a Parser.

The Lexer scans the input program and performs a lexical analysis to identify tokens and detect errors.

The Parser performs a syntax analysis and generates the object code.

The Linker combines the compiled object code with pre-compiled library files and generates executable machine code.

<https://www.learncpp.com/cpp-tutorial/introduction-to-the-compiler-linker-and-libraries/>

6.4 Example of Lexical Analysis, Tokens, Non-Tokens

A lexical element refers to a character or group of characters that may legally appear in a source file. Consider the following code that is fed to a Lexical Analyzer.

6.6 Examples of Nontokens

```
#include <stdio.h>
    int maximum(int x, int y) {
        // This will compare 2 numbers
        if (x > y)
            return x;
        else {
            return y;
        }
    }
```

6.5 Examples of Tokens Created

Lexeme	Token
int	Keyword
maximum	Identifier
(	Operator
int	Keyword
x	Identifier
,	Operator
int	Keyword
Y	Identifier
)	Operator
{	Operator
If	Keyword

6.6 Examples of Nontokens

Type	Examples
Comment	// This will compare 2 numbers
Pre-processor directive	#include < stdio.h >
Pre-processor directive	#define NUMS 8,9
Macro	NUMS
Whitespace	/n /b /t

<https://www.guru99.com/compiler-design-lexical-analysis.html>

A lexer is like a spelling checker whereas the parser checks the grammar (syntax) of programming commands.

To understand how this works, it is necessary to understand syntax. There are two usual forms of expressing syntax. The first is by means of syntax diagrams [2] the second is by a notation in the Backus-Naur Form (BNF).

6.7 Interpreters

Another way of executing algorithms is by means of an interpreter. Here each line of code is also lexically and syntactically checked before being executed (line by line). Examples of interpreters are BASIC, Python and many robot programming languages.

JAVA is interesting in that the code is first compiled into a platform independent pseudo assembly language know as bytecode which is then converted into the assembly language appropriate for the platform being used (Linux, Mac, Windows, Solaris, etc.).

For a thorough treatment of the subject of programming languages and data structures, the reader should consult appropriate texts [3].

Exercises 6−Assembler

Exercise 6.1
Break down the code LDA, R1 into the corresponding microprograms.
....

1. Read the contents from register R1 in the output register of the register bank. RNR, RES
2. Send the contents of the output register over the bus and into the main memory address register. RAS, BUS, MRW
3. Bring data from the main memory into the data register. MRW, MAR
4. Send the contents of the data register over the bus to input A of the ALU. MDR, BUS, AIN

Exercise 6.2
Continue with the example assembler programs in Table 6.2. At what point does the program sequence end?
....

After 3 Increments.

Exercise 6.3
What does "Register addressing mode" mean for the 8086 processor?
....
It means that the register is the source of an operand for an instruction.

Example

```
MOV CX, AX;copies the content of the 16-bit register AX into
the 16-bit register CX.
```

References

1. Gonnet. G.H.–*Handbook of Algorithms and Data Structures*–Addison-Wesley, 1984.
2. Blume, C., Jakob, W., Favaro, J. (1987). C Syntax Diagrams. In: *PasRo*. Springer, Berlin, Heidelberg. https://doi.org/10.1007/978-3-642-72848-8_29.
3. Tennent R.D.–*Principles of Programming Languages*–Prentice-Hall, 1981.
4. Böttcher. A. & F. Kneißl - Informatik für Ingenieure–3rd Edition, De Gruyter Oldenbourg, 2012.
5. Kernighan. B.W. & D.M. Ritchie–*The C Programming Language*–Prentice Hall, 1988.
6. XL C Enterprise Edition for AIX. XL C Language Reference Version 7.0.
7. https://www.ibm.com/docs/en/i/7.5?topic=aiccr-how-read-syntax-diagrams.
8. https://www.tutorialspoint.com/microprocessor/microprocessor_8086_addressing_modes.htm.

Applications

7

As seen in Chap. 2, many modern digital designs rely on circuit simulation and final implementation by means of programmable, rather than discrete, devices. Whichever is used, the rules of Boolean algebra apply in the same manner. Complete design of computer systems relies on extensive use of the techniques described including truth tables, Karnaugh maps, state diagrams, etc. In addition there are many advanced aspects such as electromagnetic compatibility [1], power minimisation and thermal management [2] which have not been addressed in this book.

Digital systems play an ever increasing role in our daily lives. However, that was always the case in modern automation. The rest of this chapter deals with a number of digital applications pertinent to industrial automation.

7.1 Gray Code

Simple binary buses are common within computers, but communication with machines often requires something different. When controlling the rotational position of a mechanical shaft, some means of determining the angle is necessary. For this purpose shaft encoders are often used. Because of the danger of a one track changing slightly quicker than the other, when two transitions in binary take place simultaneously, Gray code is used [3].

For the 3-bit discs, the Binary and Gray codes are shown in Fig. 7.1. It is clear to see from the truth table that two tracks change simultaneously with the binary transition from segment 1 to 2 and from 3 to 4 and again from 5 to 6 and so on. These are mechanical discs with finite precision which means there is always a possibility that one of the two

© The Author(s), under exclusive license to Springer Nature Switzerland AG 2026
G. Monkman, *Digital Electronics*, Synthesis Lectures on Electrical Engineering,
https://doi.org/10.1007/978-3-031-69726-5_7

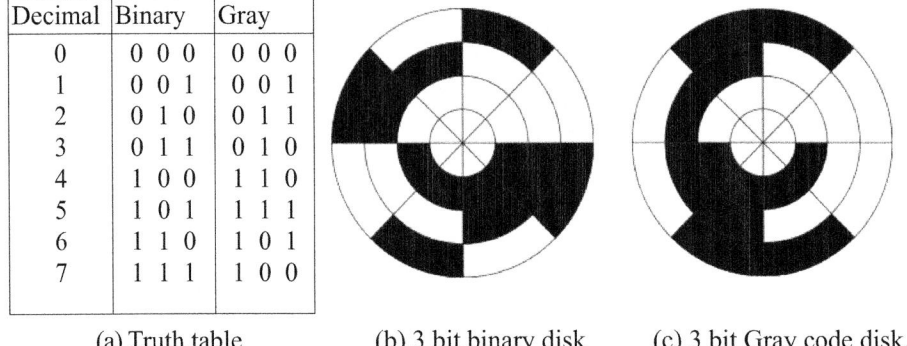

Fig. 7.1 a Truth table b 3 bit binary disk c 3 bit gray code disk

tracks will change slightly before the other. This results in a false count which may only be present for a few microseconds but that is long enough for a digital system to detect it. With Gray code, no two tracks change simultaneously meaning that this potential source of error is eliminated. It is interesting to observe that if the binary transitions 1 to 2, 3 to 4 and 5 to 6, etc. could be omitted then a Binary disc would be usable. In fact, taking every second segment in a binary disc has the same unambiguous function as a Gray code disc, though its not very efficient in that only half the code can be used.

Conversion from binary to Gray and Gray to binary can be carried out in hardware or in software (C/VHDL Boolean algebra, look-up table etc.).

7.1.1 Binary–Gray and Gray–Binary Conversions

It will be clear from Fig. 7.1 that the first two sequences are the same and that the binary and Gray MSB are always identical. A Gray code equivalent of the given binary number is computed as follows:

$G3 = B3$

$G2 = B3 \oplus B2$

$G1 = B2 \oplus B1$

$G0 = B1 \oplus B0$

A binary to Gray code converter can be implemented using XOR gates. For n inputs, n-1 gates are required. As shown in the image of Fig. 7.2, for 4 inputs, 3 XOR gates are used:

Binary code equivalent of the given Gray number is computed as follows:

7.1 Gray Code

Fig. 7.2 Binary to gray conversion

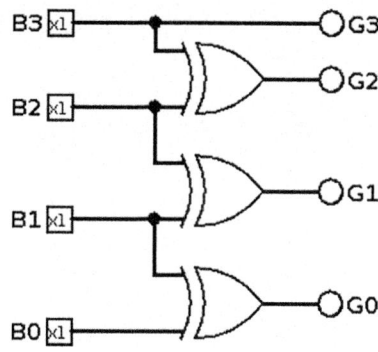

Fig. 7.3 Gray to binary conversion

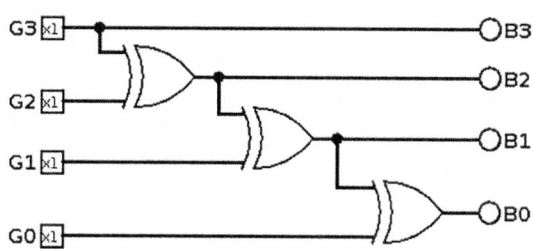

$B3 = G3$, $B2 = B3 \oplus G2$, $B1 = B2 \oplus G1$, $B0 = B1 \oplus G0$

A Gray to binary code converter can also be implemented using XOR gates. Again, for n inputs, n-1 gates are required. As shown in the image of Fig. 7.3, for 4 inputs, 3 XOR gates are used:

However, with the above Gray to binary converter, the problem of propagation delay means that the original problem is not fully solved. Unfortunately, the XOR cascade circuit still has problems due to propagation delay in the gates. Output B3 will appear before B2 and B2 before B1 and so on. To avoid this problem, the outputs must be passed to a parallel series of latches before the resulting binary code is clocked to its final destination as soon as all outputs are stable.

An alternative is to read the code verbatim and carry out the conversion in software. Two simple algorithms are shown below:

Binary to Gray (MSB is bit 0, b is Binary, g is Gray Code):

```
if b[i-1] = 1
    g[i] = not b[i]
else
    g[i] = b[i]
```

Or:

```
g = b xor (b logically right shifted 1 time)
```

Gray to Binary (MSB is bit 0, b is Binary, g is Gray Code):

```
b[0] = g[0]

for other bits:
b[i] = g[i] xor b[i-1]
```

The above algorithms may also be coded in C or VHDL as required:

Equivalent in C

```
integer
gray_encode(integer n)
{
    return n ^ (n >> 1);
}
integer
gray_decode(integer n)
{
    integer p;
    p = n;
    while (n >>= 1) {
        p ^= n;
    }
    return p;
}
```

7.1.2 Incremental Counters

As a cost-effective alternative to absolute encoders such as Gray code discs, incremental encoders are often used. These comprise a disc with equally spaced concentric holes through which light may pass. An appropriate optoelectronic circuit then detects whether light passes through or not. Rotation of the disc produces a series of pulses, the number of which is proportional to the rotational angle of the disc. A practical implementation using an up/down counter circuit, as seen in Chap. 3, for an incremental encoder is shown in Fig. 7.4.

Counters are not limited to simple binary or BCD sequences but can be made to produce almost any codes depending on the additional logic employed.

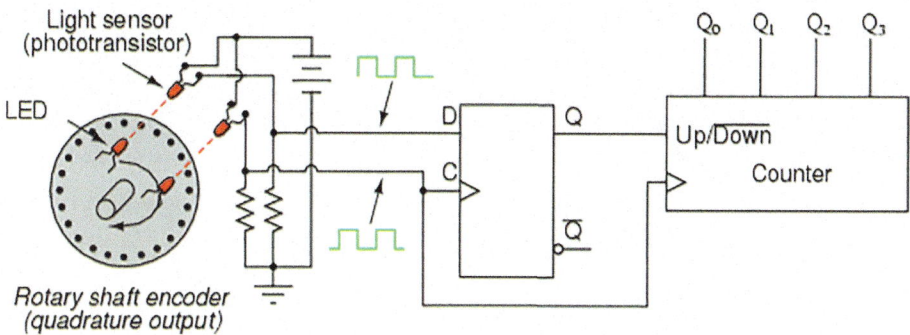

Fig. 7.4 Application for up/down counter (Courtesy: all-about-circuits)

7.2 Parallel Bus Systems

Having hitherto concentrated on the individual digital components and modules necessary to build digital equipment, this chapter deals with connecting them together in order to produce complete systems. To these ends there must be some means of connecting the individual modules together. This is usually achieved using parallel bus systems or serial bus systems.

Although now considered almost obsolete for external use, parallel bus systems remain the main internal communication means between computer modules. There are essentially three bus systems linking all the architecture of a computer together. The *control bus* is used by the CPU to tell the memory it wishes to store or retrieve data to and from memory along the *data bus*. For this purpose the required address of the memory location in question must be provided by the CPU along the *address bus*. The same applies to data sent to the outputs and received from the inputs by means of the *data bus*. This is illustrated in Fig. 7.5.

Attaching a multi-channel oscilloscope to such a bus would reveal a number of signals (8 for an 8-bit bus) all synchronized via a clock signal (Fig. 7.6).

A typical example of a simple parallel bus structure linking a Texas Instruments TDS9092 processor with the outside world, employs a number of bus drivers for inputs and outputs. The processor accesses the desired input or output module by means of a 74HC138 decoder. Binary values on the address lines 0, 1 and 2 can allow up to 8 outputs on the decoder to be selected. These, in turn, enable the appropriate driver (74HC541 for inputs and 74HC574 for outputs).

Normally, such bus drivers have tri-state outputs to allow the connections to "float" in order to avoid conflict between inputs and outputs accessing the same bus lines. Where

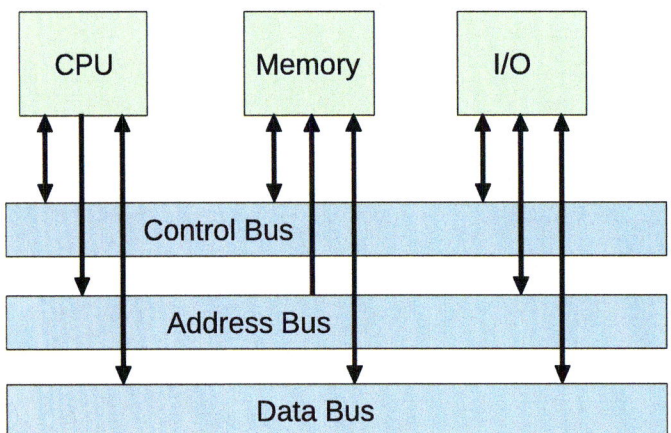

Fig. 7.5 Computer internal bus system

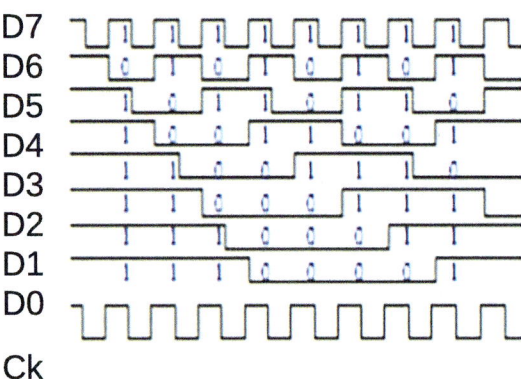

Fig. 7.6 Typical 8-bit bus signal lines

PLCs are concerned, these are usually augmented with a degree of galvanic separation in the form of relays or optocouplers (Fig. 7.7).

There are a number of parallel bus systems which are still commonly used. Error detection in serial bus systems is often achieved using parity or checksum. With parallel bus systems it is more difficult without adding additional lines. However, taking the usual BCD and adding 3 to it gives Excess-3 (XS3) code. XS3 is often used as it does not contain the combinations 0000 or 1111 in the 4-bit example. Given a simple 4-bit line, 0000 would normally represent an absent connection whereas 1111 suggests a short circuit on all lines. Both are common faults with parallel bus systems (Table 7.1).

As seen earlier, Gray code is another often used code for machine interfaces. With Gray code, only one bit changes at any transition. This is important where mechanical devices are employed. With BCD it is possible for two bits to change simultaneously. For

7.2 Parallel Bus Systems

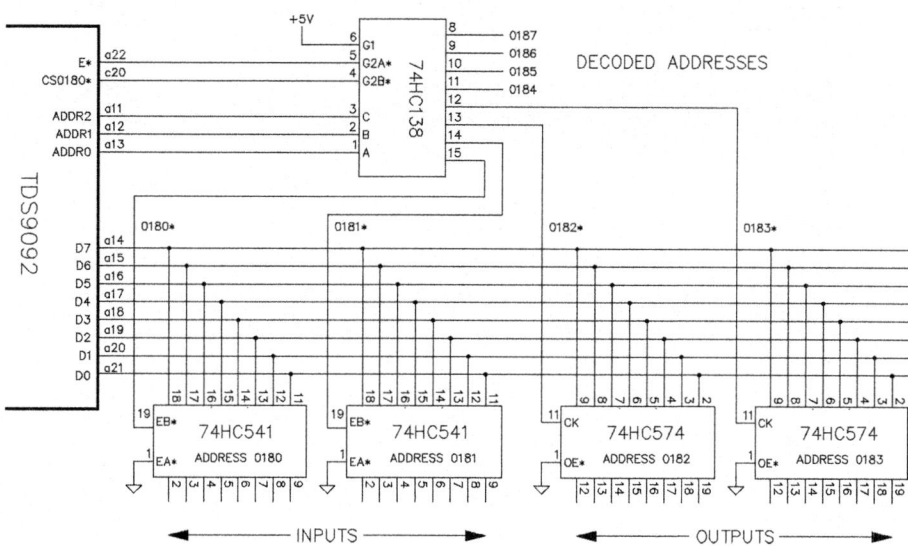

Fig. 7.7 Microcontroller connected to inputs and outputs via a parallel bus system (Courtesy: Triangle digital)

Table 7.1 Examples of 4-bit BCD, XS3 and Gray codes

Decimal	BCD (8-4-2-1)	XS3 (BCD + 0011)	Gray code
0	0 0 0 0	0 0 1 1	0 0 0 0
1	0 0 0 1	0 1 0 0	0 0 0 1
2	0 0 1 0	0 1 0 1	0 0 1 1
3	0 0 1 1	0 1 1 0	0 0 1 0
4	0 1 0 0	0 1 1 1	0 1 1 0
5	0 1 0 1	1 0 0 0	0 1 1 1
6	0 1 1 0	1 0 0 1	0 1 0 1
7	0 1 1 1	1 0 1 0	0 1 0 0
8	1 0 0 0	1 0 1 1	1 1 0 0
9	1 0 0 1	1 1 0 0	1 1 0 1

example, should the LSB change from 1 to 0 before the other bits in a 3 to 4 transition in a BCD system, then the computer would think it is at position 2 instead of position 4.

Of course all bus systems can be extended to 8, 16, 32 or 64 bit depending on requirements. An example of a commonly used 8-bit code is the ASCII code seen in Chap. 2.

7.2.1 7-Segment Display

One of the commonest encoders is one which converts BCD to 7-segment display format. Whether LED, liquid crystal or plasma, 7-segment displays are ubiquitous in modern electronics. As shown in Fig. 7.8, there are 7 selectable segments (A, B, C, D, E, F and G) plus a decimal point (DP).

There are two types of 7-segment display, common cathode and common anode. In the common cathode version, all LED cathodes are connected together and must be grounded. Each LED is then driven by a logic 1 (as depicted in Table 7.2). For the common anode version, all LED anodes are connected together and to + Vcc. Each LED cathode is then selected by a logic 0. Several integrated circuits suitable for this task are available, for example: TTL 7447 and CMOS 4511 <https://www.electronics-tutorials.ws/blog/7-segment-display-tutorial.html>.

Normally a current limiting resistor must be inserted between the output of the driver and the LED. The value of the resistor depends on the LED current and voltage drop (which is usually given by the manufacturer).

7.3 Error Detection

Error detection plays an important role in the transfer of data over bus systems. The use of XS3 and Gray code for single lines of data were described earlier. For larger amounts of data, error detection becomes more involved.

Simple check systems such as those associated with parallel bus systems using XS3 are not particularly accurate. Many modern bus systems are serial, and the data must be received and stored before it can be checked for accuracy. For these purposes there are a number of techniques.

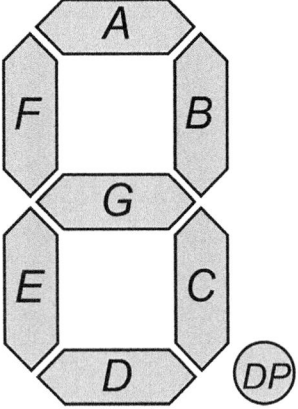

Fig. 7.8 7-segment display

7.3 Error Detection

Table 7.2 Truth table for BCD to 7-segment conversion

Decimal	A	B	C	D	E	F	G
0	1	1	1	1	1	1	0
1	0	1	1	0	0	0	0
2	1	1	0	1	1	0	0
3	1	1	1	1	0	0	1
4	0	1	1	0	0	1	1
5	1	0	1	1	0	1	1
6	1	0	1	1	1	1	1
7	1	1	1	0	0	0	0
8	1	1	1	1	1	1	1
9	1	1	1	0	0	1	1

7.3.1 Parity Checking

An ASCII character can be encoded as seven bits where each character is stored as an eight-bit byte and the additional bit used as a parity check.

In even parity the parity bit is used to ensure that the total number of 1-bits in the byte is always even. On the other hand, in odd parity the parity bit is employed to ensure that the total number of 1-bits is odd. The sender sets the parity before transmission and the receiver checks it on receipt, and requests a resend if necessary. If the parity bit is found to be wrong then the character has been corrupted during transmission.

Although a parity check will detect a single bit error in a character (a single zero altered to a one or vice versa) it will not recognise a corruption in which two simultaneous errors have cancelled each other out. Thus, a more sophisticated check can be accomplished by dealing with a complete block of data each time. In addition to calculating the parity bit for each character (horizontal parity), the parity for all equivalent bits in the block is calculated (vertical parity). This forms a **Block Check Character (BCC)** which is illustrated in Table 7.3.

Although two-coordinate parity checking can be useful for data transmission, errors can still go undetected and it does carry a considerable overhead in terms of the number of additional parity bits which must be transmitted.

Checksums

In modern computer communications environments an interesting alternative to simple parity bits is the **Cyclic Redundancy Checksum (CRC)**. In this context the complete block of data is treated as a single binary integer. An integer division by some constant is carried out on this, hence leaving a quotient and a remainder. The remainder is transmitted along with the data block and is compared to the remainder obtained by the receiver on

Table 7.3 Block check example [http://www.oocities.org/ernestwu/networking/parity.htm]

	Data transmitted		Data Received
A	10000010	Bits 2 & 4 changed	11100010
B	10000100		10000100
C	10000111		10000111
D	10001000		10001000
E	10001011		10001011
F	10001101		10001101
G	10001110	Bits 7 & 8 changed	10001101
Space	01000001		01000001
H	10010000		10010000
I	10010011		10010011
J	10010101		10010101
K	10010110		10010110
BCC	11000000	Bits 2, 4, 7 & 8 changed	10100011

performing the same calculation. If they match exactly then transfer of the information must have occurred without corruption. [http://www.oocities.org/ernestwu/networking/parity.htm].

In Unix based systems a checksum can be generated for a given file:

Syntax: cksum [File …].

Example

$ cksum test.txt

4038471504 75 test.txt

where "4038471504" represents the checksum value and "75" represents the file size of test.txt

Another extremely robust form of encoding for transmission uses Manchester encoding. Compared with the simple 1 in 8 safety provided by parity. Manchester codes deliver a 12 from 16 safety ratio.

7.3.2 Differential Manchester Encoding

There are two possible and equivalent definitions for the Manchester code, as given below and also depicted in Fig. 7.8.

1. In the code definition according to G.E. Thomas, a falling edge means a logic one and a rising edge a logic zero. This definition is also called Biphase-L or Manchester-II.
2. In the code definition according to IEEE 802.3, as used for 10-MBit/s Ethernet, a falling edge means a logic zero and a rising edge means a logic one.

In each case, there is at least one edge per bit from which the clock signal can be derived. The Manchester code is self-synchronizing and independent of the DC voltage level. To inform the receiver how a logic 1 is coded in the signal, a header (preamble) can be sent at the beginning of a data transmission.

Instead of sending the binary signal itself, Manchester encoding converts transitions in the binary into unique 4-bit codes. There are four possibilities: 0 to 0; 0 to 1; 1 to 0 and 1 to 1.

Advantages

Each binary transition results in 4 bits, although there are 16 possibilities, i.e. 12 are not used and so indicative of a transmission error. This is an error detection likelihood of 3 to 4 as compared with simply parity which gives 1 in 8.

The clock signal can be derived from the code itself. An additional clock generator is not required.

There is no DC component. Therefore, it is possible to transmit the signal sequence via pulse transformers, optoisolators etc. so guaranteeing galvanic isolation.

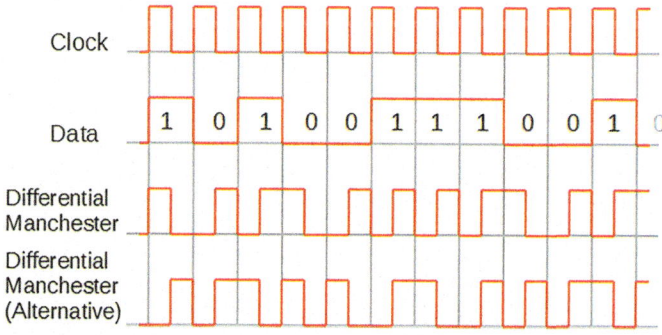

Fig. 7.9 Manchester encoding (Courtesy: wikimedia.commons.org)

Disadvantages

One disadvantage of Manchester coding is that the bandwidth required for data transmission is twice as high as with simple binary coding (e.g. Non Return to Zero, NRZ code). The reason for this is that two signals are needed to encode one bit. The bit rate (in the case of a two-valued signal) is therefore only half as large as the baud rate.

Complete design of advanced desktop and laptop computers of the sort we use every day is beyond the scope of a first-year undergraduate course. However, the design and construction of smaller digital systems should be possible at this level. Industrial examples include dedicated microcontrollers and Programmable Logic Controllers (PLC).

7.3.3 Programmable Logic Controllers (PLC)

PLCs are stand alone computers intended for controlling automated equipment. They differ in hardware from normal computers in that they additionally have buffered outputs and inputs. That means the I/O are galvanically separated from the equipment to be controlled. Electrical isolation is essential in industrial environments where noise is present, particularly where high voltages and heavy currents are to be switched.

For inputs, optical isolation is the most common method. For outputs electromechanical relays are often used. Other methods, such as inductive, acoustic and even thermal are also occasionally employed.

From a software point of view, there are no operating systems like Windows or Linux (though Linux is often used in embedded systems). Although offline programming may be carried out from a separate computer, the PLC program itself looks very similar to assembly code.

As a visual aid, this code may be converted to "contact plan", "digital circuit" or even "structured text" formats as required. One can even program in one of these alternatives, but the basic functions are converted to PLC programming commands for execution and simulation.

There are 5 types of PLC programming languages defined by the **IEC 61131-3** standard:

Ladder Logic Diagram (LD)—Contact plans as used by electrician of the old school.

Instruction List (IL)—similar to assembler.

Function Block Diagram (FBD)—logic block diagram form.

Structured Text (ST)—like modern computer programming.

Sequential Function Charts (SFC)–Grafcet flow charts (**DIN EN60848**).

A ladder diagram or contact plan is basically an electrical circuit diagram in the traditional form used by industrial electricians. To implement an AND function two switches in series are needed.

7.3 Error Detection

Fig. 7.10 Contact plan for an AND function

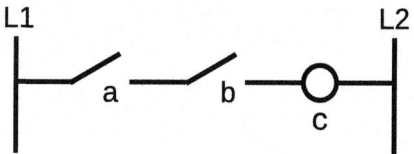

In Fig. 7.10, L1 and L2 are simply the two sides of a power supply, for example + 24V and ground.

The following PLC code instruction list (IL) for the AND function should be very easy to understand by now:

```
PROGRAM PLC_PRG
    VAR
        a: BOOL;
        b: BOOL;
        c: BOOL;
    END_VAR

    LD    a
    AND   b
    ST    c
```

The first part simply defines the variables (as in the programming language C). The second part is the program:

1. Load the input a into the accumulator.
2. The contents of the accumulator (a) AND b is written into the accumulator.
3. The contents of the accumulator are stored in output c.

As a function block diagram (FBD) this is simply a logic gate as implemented in a simulator (Fig. 7.11):

This is equivalent to c: = a AND b; in structured text (ST) form.

Grafcet is a graphical technique ideally suited to automation systems.

Steps are represented in sequential function charts such as Grafcet as squares and are given a unique alphanumeric labelling within the square. Several actions, displayed as a rectangle and box, can be attached to a step. Inputs are given at transitions between

Fig. 7.11 Function block diagram of the AND function

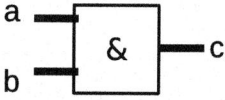

Fig. 7.12 Grafcet representation of the AND function

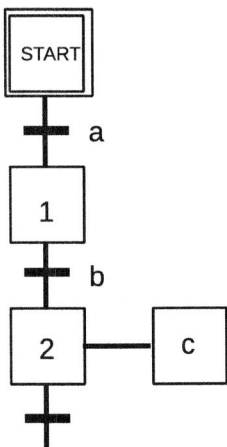

labelled steps. Outputs are also represented by quadratic boxes to the right. An example is shown in Fig. 7.12.

In most PLC systems it is possible to switch between representations, though some older systems are limited to one or two possibilities. Whichever representation is used, it should be clear that, for execution, such code will be transformed into machine code with a similar structure and executed at processor level in binary.

Exercises 7 – Applications

Exercise 7.1
What is the problem with hardware digital gray to binary converters and what other hardware should it be equipped with to avoid this problem?
....

Since gray-to-binary converters consist of cascades of EXOR gates, not all outputs are present at the same time. Therefore, a bank of latches would be necessary so that all outputs can be clocked out at the same time.

Exercise 7.2
What are the advantages and disadvantages of Manchester coding?
....

Advantage: only 4 out of 16 code combinations are valid - very robust.

Disadvantage: doubled bandwidth.

Exercise 7.3

Which 5 components are essential for every PLC system?

....

Control unit, memory, input unit, output unit, timing units (counters & timers).

Exercise 7.4

Display the following PLC instruction list (IL) in the form of a functional block diagram (FBD).

```
LD    NOT  a
AND   NOT  b
ST    x
```

(Given that a, b, x are declared as type BOOL).

....

References

1. Dawson, J.F., Ganley, M.D., Robinson, M.P., Marvin, A.C., Porter, S.J.—Design For Electromagnetic Compatibility. In: Huang, G.Q. (eds) *Design for X.*—Springer, Dordrecht, 1996. https://doi.org/10.1007/978-94-011-3985-4_14.
2. Benson. M.-*The Art of Software Thermal Management for Embedded Systems.*—Springer New York 2014.
3. Bentley. J.P.-*Principles of Measurement Systems*—John Wiley, 1988.

Internet Sites

4. http://www.allaboutcircuits.com/textbook/digital/chpt-11/synchronous-counters/.
5. http://electronicsgurukulam.blogspot.de/2012/08/binary-to-gray-code-converter-gray-code.html.
6. https://www.electronics-tutorials.ws/blog/7-segment-display-tutorial.html.
7. https://www.gnu.org/software/coreutils/manual/html_node/cksum-invocation.html.
8. https://www.realdigital.org/doc/fc26cf6e35d2a61c6e2871dd9be9e21a.
9. http://rosettacode.org/wiki/Gray_code.
10. https://www.triangledigital.org.uk.
11. https://www.solisplc.com/blog/plc-programming-languages.
12. http://8051-microcontrollers.blogspot.de/2014/11/number-systems-binary-coded-decimal-and.html#.VaohkKY27VY.

A Short History of Computers

History is particularly interesting when it concerns a subject with which the reader is already familiar with. However, one must be careful as many historical texts have national tendencies. The waters are further muddied when such things as official secrets acts play a role. This is particularly important with modern technology because of its inherent military relevance.

Despite such hindrances, this text attempts to provide the reader with an overview of events leading to the modern computer as we know it. How the first devices and systems were invented is very relevant to the functional comprehension of modern digital elements. Not only because early devices were larger and more observable than their modern counterparts but also because the needs of the time are embedded in their conception and realization. This helps to understand the how and why of such developments.

As explained in Chap. 5, all computer systems require a calculating unit, a memory and some form of input and output. The first calculating machines were purely mechanical. Input was by means of mechanical settings and outputs as numbers, also displayed mechanically. The main problem with such devices was the inherent difficulty in storing information. Pre-programmed cards and tapes went some way in achieving very primitive programming but temporary data storage proved much more difficult.

With the advent of electricity, the ability to temporarily store information (memories) for both temporary and permanent use became much easier to implement. In addition to those dealt with in Chap. 3, many physical techniques including electro-mechanical, magnetic, electrostatic and even acoustic and optical have all contributed to the development of modern memory systems.

The next great breakthrough was the microprocessor. This condensed the traditional collection of discrete ICs (usually spread over several plug-in PC boards) onto one single main chip and a few interface chips (maximum one PCB).

8.1 A Short History of Calculating Machines

The very first "computers" were simple calculating machines like the abacus. The first step in the extension of the abacus for automatic calculation was taken in 1623, when the German astronomer Wilhelm Schickard (1592–1635) constructed his "calculating clock", as shown in Fig. 8.1. This machine was capable of executing addition and subtraction.

The possibility of mechanising arithmetic was first publicly demonstrated in 1642, when Blaise Pascal (1623–1662), totally unaware of the achievements of his predecessor Schickard, constructed his celebrated "Pascaline". The principal characteristic of Pascal's machine was its facility for automatic carrying. This was achieved by the use of a series

Fig. 8.1 Schickard's "Calculating clock" (Courtesy: Herbert Klaeren). *Source* https://upload.wikimedia.org/wikipedia/commons/1/17/CHM_Artifacts_Pascaline_calculator_%282632674049%29.jpg

8.1 A Short History of Calculating Machines

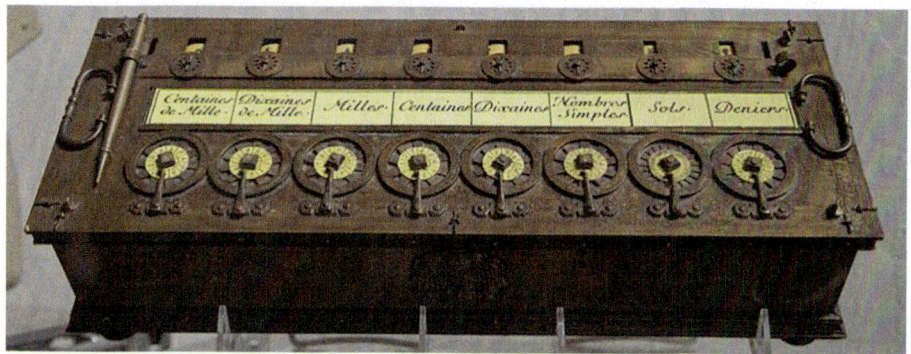

Fig. 8.2 The "Pascaline" (Courtesy: Martin Wichary). *Source* https://upload.wikimedia.org/wikipedia/commons/5/5a/Schickardmaschine.jpg

of toothed wheels, each numbered from 0 to 9, linked (by weighted ratchets) in such a way that when one wheel completed a revolution the next wheel advanced by one step. The prototype had five wheels, and so could handle five-digit numbers; later versions had six or eight wheels (Fig. 8.2).

Both Gottfried Wilhelm Leibniz and Isaac Newton came upon the idea of differential calculus almost simultaneously. Though arguments raged for many years as to who was the first. An excellent example of nationalist claims! What is certain, is the fact that Leibniz invented a machine for carrying out all four fundamental arithmetical functions (addition, subtraction, multiplication and division) by purely mechanical means, though the abacus can also achieve this [6] (Fig. 8.3).

Conceived in 1673 but only built in 1694, unaware of Schickard's work and borrowing nothing from Pascal, Leibniz devised mechanisms which would carry out multiplication and division by means of successive additions and subtractions.

Subsequent generations of inventors gradually moved away from the basic ideas of Schickard, Pascal and Leibniz. Their machines brought a number of detailed improvements with them. Amongst these were those invented by: the Italian Giovanni Poleni (1709), which was distinguished by the use of gears with variable numbers of teeth; the Austrian Antonius Braun (1727); the German Jacob Leupold (1727), improved by Antonius Braun in 1728 and built in 1750 by a mechanician called Vayringe; the German Philipp Matthaus Hahn developed in 1770, of which a series of machines were designed between 1774 and 1820; the Englishman Lord Stanhope whose two calculators were developed around 1775 to 1777; the German Johann Hellfried Muller (1782–1784); etc. The first major advance after Leibniz's invention was made by the French engineer and industrialist Charles-Xavier Thomas of Colmar, director of a Paris insurance company, who in 1820 invented a calculator which he named the *Arithmometer* [1] (Fig. 8.4).

Fig. 8.3 Leibniz's arithmetical machine (Courtesy: Wikimedia commons). *Source* https://commons.wikimedia.org/wiki/Category:Leibniz_calculator#/media/File:Leibniz_Rechenmaschine_(1690).jpg

Developed in 1822 and constantly improved during the following decades, the machine was conceived on similar lines to that of Leibniz. The "Leibniz Gears" were now fixed in position instead of sliding horizontally, the pinion which engaged each of them being made to rotate about its axis.

Finally, Tchebishev's calculator of 1882, notable for its epicyclic gear mechanism for carrying numbers, as well as a component which automatically shifted the carriage during multiplication, by which this operation became effectively automatic (Fig. 8.5).

The slow performance of early calculating machines was largely due to the data entry method which still required the close attention of the operator. The numeric keyboard, made in the middle of the nineteenth century, was a breakthrough resulting from the development of the typewriter. To enter each digit, it is sufficient to press once with a single finger on the appropriate key which automatically returns to its starting position once released. This form of data entry was simpler, quicker, more precise and very efficient.

Decisive developments in this area occurred between 1885 and 1893, when the American William S. Burroughs invented and perfected his Adding and Listing Machine, shown in Fig. 8.6a). This was the first mechanical calculator with keys and a printer which was also practical, reliable, robust and perfectly adapted to the requirements of the banking

8.1 A Short History of Calculating Machines

Fig. 8.4 Arithometer (Courtesy: Smithsonian). *Source* https://americanhistory.si.edu/collections/nmah_690457

system and commercial operations of the time. The complete solution to the printing problem for adding machines was also achieved at almost the same time (1889–1893) by Dorr E. Felt who had invented the Comptometer, depicted in Fig. 8.6b).

For a comprehensive review of mechanical calculating machines, the interested reader should consult the more extensive text written by George Chase [1].

These were the first calculating machines but none of them were really programmable. The first real programmable computer may be traced back to Charles Babbage's difference Engine as early as 1822 (Fig. 8.7).

Charles Babbage (1791–1871), computer pioneer, designed the first automatic computing engines. He invented several computers but failed to completely build them. The first complete Babbage Engine was finally built (from the original plans) in London in 2002—153 years after it was designed. Difference Engine No. 2, built faithfully to the original drawings, consists of 8,000 parts, weighs five tons, and measures almost 4 m in length.

Although in Babbage's lifetime his engine was never built, the idea of programming was conceived by an amateur mathematician Lady Ada Lovelace Byron. Although such

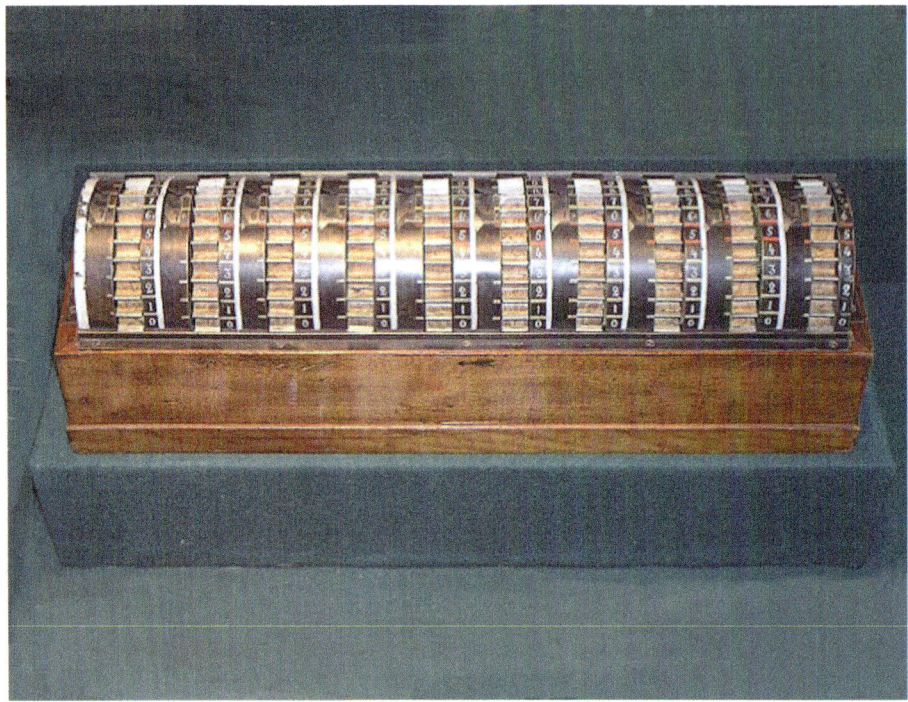

Fig. 8.5 Chebyshev calculator (Courtesy: Andrew Butko). *Source* https://upload.wikimedia.org/wikipedia/commons/e/ee/%D0%90%D1%80%D0%B8%D1%84%D0%BC%D0%BE%D0%BC%D0%B5%D1%82%D1%80.jpg

programs could never be executed on a computer that didn't exist, the algorithm for the first ever program, to calculate Bernoulli numbers, was written [4].

8.2 Memory

The ability to store data is essential for any computer system which transcends the operation of basic pocket calculators.

Early systems were inevitably purely mechanical. The first was the punched card—originally invented by a weaver, Joseph-Marie Jaquard in 1801 for programming looms. Its introduction to calculation machines came with Herman Holerith of IBM in 1890. A concept used for many years for storing music, the first known use of punched paper tape in calculating technology was in 1846 by Alexander Bain–the inventor of the

8.3 Memory Cells

(a) (b)

Fig. 8.6 a Burroughs. b Felt and Tarrant Comptometer Model H (Courtesy: Smithsonian). *Source* https://americanhistory.si.edu/collections/nmah_690457

fax machine and the electric printing telegraph. Each row on the tape represents one character, but as the tape is continuous it can store significantly more data than punched cards (Fig. 8.8).

Punched card and tapes have one significant disadvantage: data can only be stored and retrieved in a fixed sequential manner. An absolute necessity for modern computers is the ability to randomly access individual memory cells.

8.3 Memory Cells

As explained in Chap. 3, computer static memory cells consist of bistable elements. With the appropriate cross-coupling they can be made to oscillate—astable devices, sometimes known as multivibrators.

The first electrical bistable devices were made using electromagnetic relays (Fig. 8.9).

Purely electronic bistable and astable devices were first created using thermionic valves (tubes) (Fig. 8.10).

Fig. 8.7 Part of George Babage's difference engine (Courtesy: Marc Smith). *Source* https://www.flickr.com/photos/49503165485@N01/3616162990

Although like single transistors, single triodes were available, more common was the employment of double triodes. These were already widespread because of their use in audio push–pull output circuits of the time. Much miniaturization was seen between the first valves and the those manufactured during 1960s. With the advent of the bipolar transistor, miniaturization took a further step forward (Fig. 8.11).

This became the usual circuit until integrated circuits came along. Incidentally, the valves in the circuit shown in the triode astable could be directly replaced with field effect transistors.

In fact, almost any two active electronic components can be used to build such circuits. Figure 8.12 shows an astable multivibrator made from two SCRs (Thyristors). C1/R1 and

8.3 Memory Cells

Fig. 8.8 A programmable Jaquard loom (Courtesy: Hélène Rival). *Source* https://upload.wikimedia.org/wikipedia/commons/1/16/03_Metier_jaquard_haut.jpg

C2/R2 determine the switching times <http://chemelec.com/Projects/Misc-Circuits/SCR-Flipflop.png>.

Instead of free running oscillation, in computers it is more common to have controlled switching of the two binary states. The astable circuit shown in Fig. 8.11 is useful as a clock generator but to make a controllable bistable element the capacitors must be removed to make it stable as depicted in Fig. 8.13.

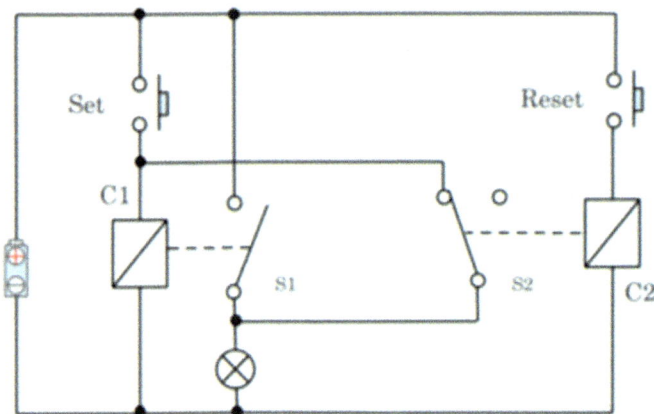

Fig. 8.9 Bistable using relays (Courtesy: Homofaciens). *Source* https://homofaciens.de/bilder/technik/relay_019.htm

Set = 0 & Reset = 0:   C1 = off, C2 = off, S1 = 0, S2 = 1, Lamp = off

Set = 1 & Reset = 0:   C1 = on,   C2 = off, S1 = 1, S2 = 1, Lamp = on

Set = 1 & Reset = 1:   C1 = on,   C2 = on,   S1 = 1, S2 = 0, Lamp = on

Set = 0 & Reset = 1:   C1 = off, C2 = on,   S1 = 0, S2 = 0, Lamp = off

Set = 0 & Reset = 0:   C1 = off, C2 = off, S1 = 0, S2 = 1, Lamp = off

The ability to set and reset must also be incorporated as in the first relay bistable example of Fig. 8.9.

As computers became more common (and smaller) integrated circuits took over from discrete transistor circuits. Bistable circuits today comprise NOR or NAND gate configurations with high degrees of circuit complexity.

8.4 Non-Volatile Memory

All the above configurations suffer from a major drawback-they are volatile! This means that when they are switched off they lose their memories. One of the first ways of making non-volatile memories was by using polarized, or *Carpenter* relays (Fig. 8.14).

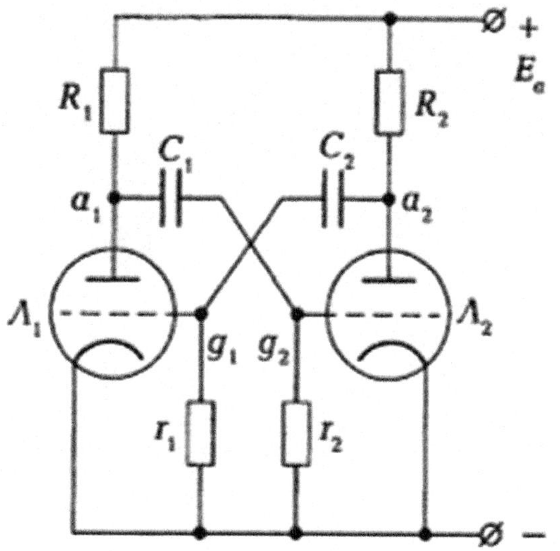

Fig. 8.10 Astable multivibrator using triodes [8]

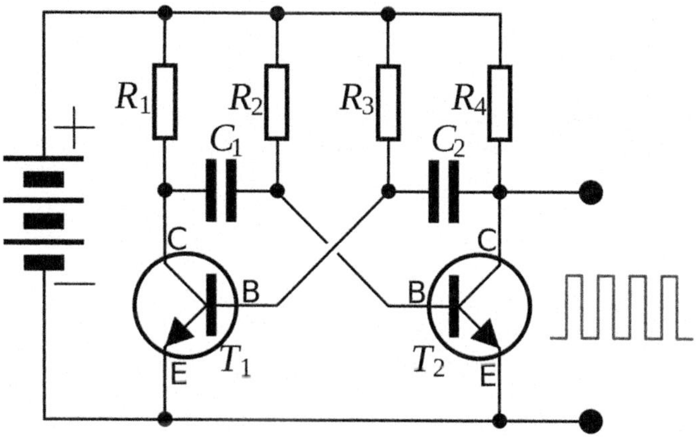

Fig. 8.11 Astable device using bipolar transistors (Courtesy: Wikimedia commons). *Source* https://upload.wikimedia.org/wikipedia/commons/3/39/Astable_multivibrator.png

A current pulse through the windings in one direction causes the armature to make the contacts at one side. Due to the permanent magnet the armature remains in position on cessation of the current. However, electromechanical components were considered too large and bulky for computers so during the 1950s an alternative was sought.

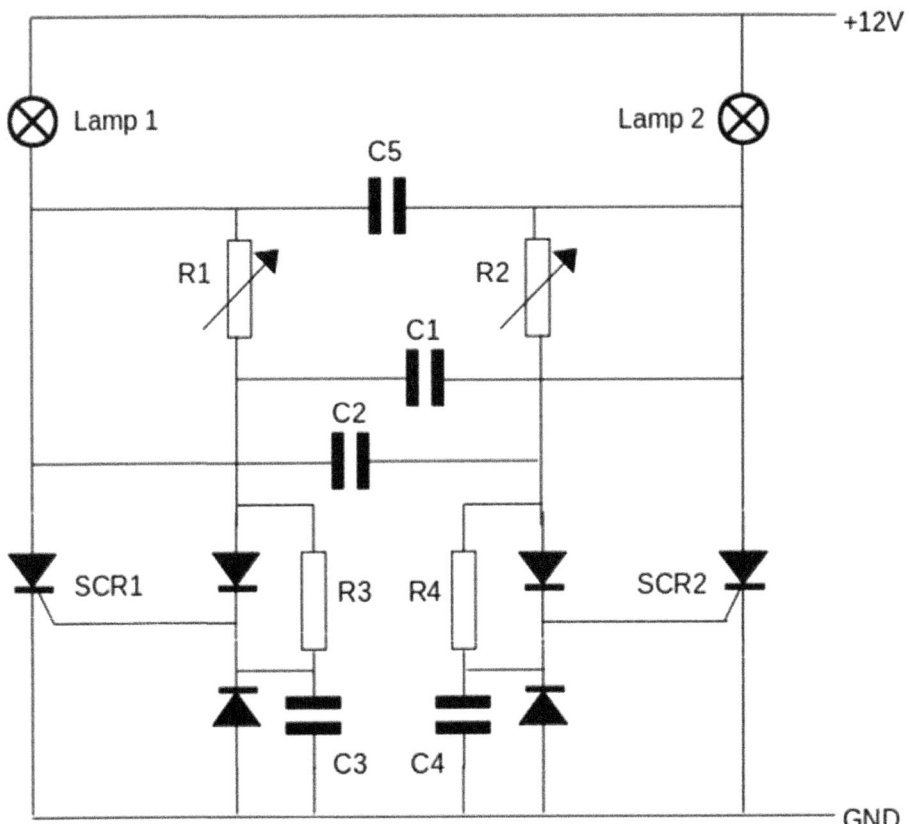

Fig. 8.12 Astable circuit using thyristors. *Source* http://chemelec.com/Projects/Misc-Circuits/SCR-Flipflop.png

8.5 Memory Arrays

The history concerning the many attempts to store data makes very interesting reading.

If a signal pulse can be delayed during propagation then so can subsequent pulses be delayed. This means that a series of pulses can be passed around in a loop like data in a shift register (also a form of serial data storage). Before the advent of digital electronics, as we know it today, many such ideas were investigated. Radio frequency delay lines have been around since the 1920s [3]. Radio waves can travel at the speed of light but acoustic waves are considerably slower (depending on the propagation medium). Acoustic signals sent through a mercury medium can be picked up at the other end and fed back (rolled

8.5 Memory Arrays

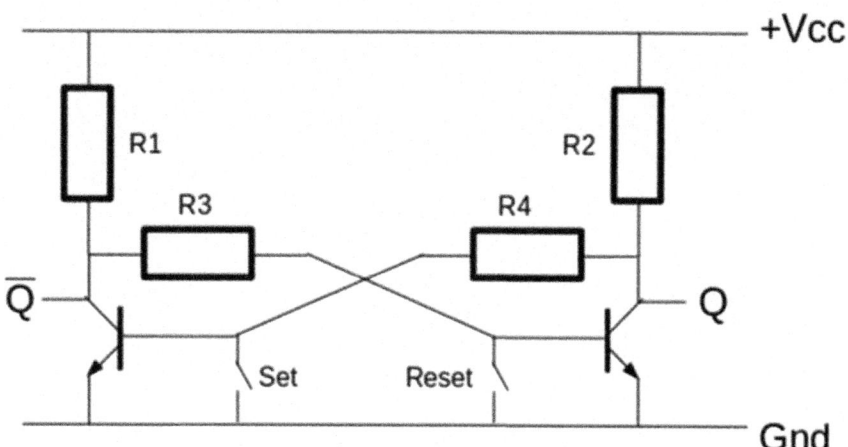

Fig. 8.13 Transistor bistable circuit with set and reset inputs

Fig. 8.14 Polarized relay

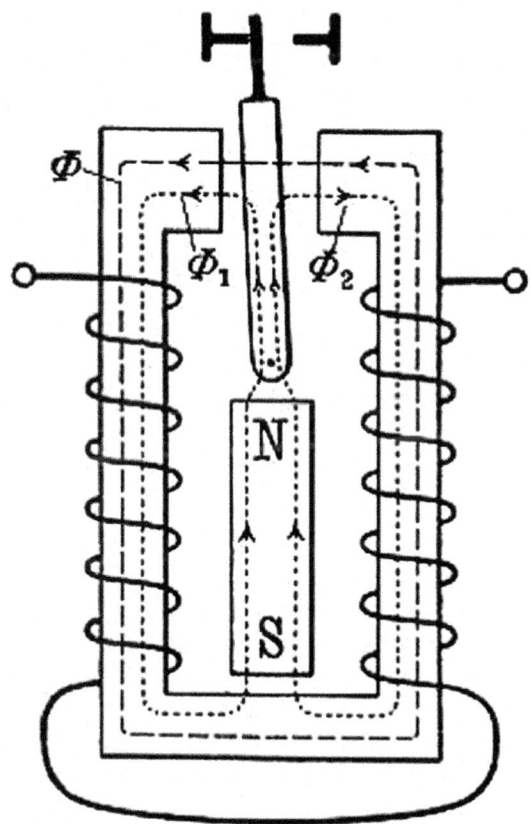

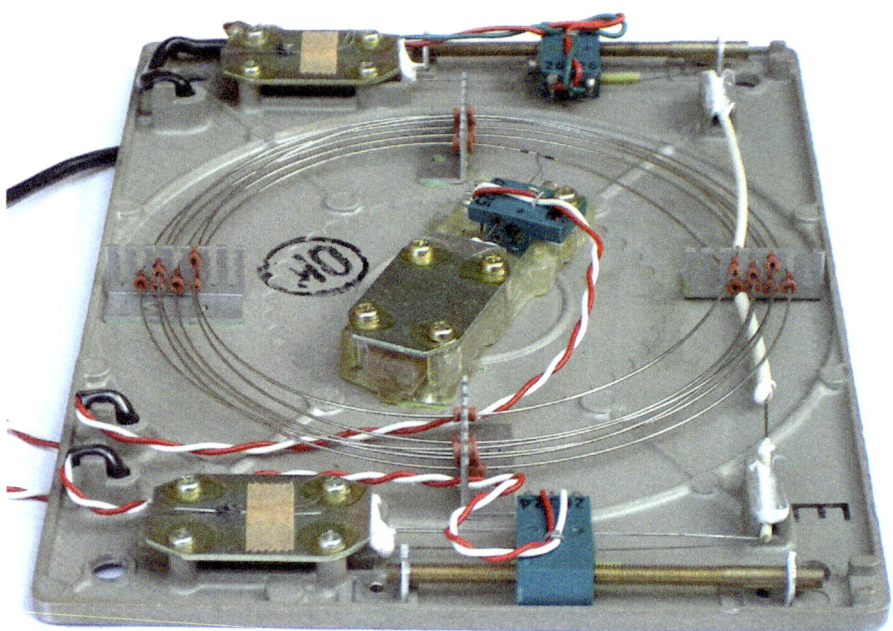

Fig. 8.15 Magnetostrictive delay storage (Courtesy: Coronium). *Source* https://upload.wikimedia.org/wikipedia/commons/e/ef/Torsion_wire_delay_line.jpg

as in modern shift registers). The bit train sent in this loop is the stored data. A more modern version uses the magnetostrictive properties of coiled nickel wire, as shown in Fig. 8.15, but the principle is the same.

In 1947 Manchester Universities, Freddie Williams and Tom Kilburn developed a high-speed, entirely electronic memory using a cathode ray tube to store bits as dots on the screen. The dots were detected as electrical charge on the screen surface (opto-sensors would today also be viable alternatives). Because the screen phosphor has a finite persistence, the dots must be refreshed periodically. Using 2D materials comprising photonics and electrets, these techniques are again topics of research interest [9]. Since the advent of the compact disc (CD), permanent (and sometimes rewritable) optical storage has been in every day use.

However, the first widely used non-mechanical non-volatile memories were electromagnetic-but without moving parts. In core memory, small ring-shaped magnets-the cores-are threaded by two crossed wires, X and Y, to make a matrix known as a plane. When one X and one Y wire are powered, a magnetic field is generated at a 45-degree angle to the wires (Fig. 8.16).

8.5 Memory Arrays

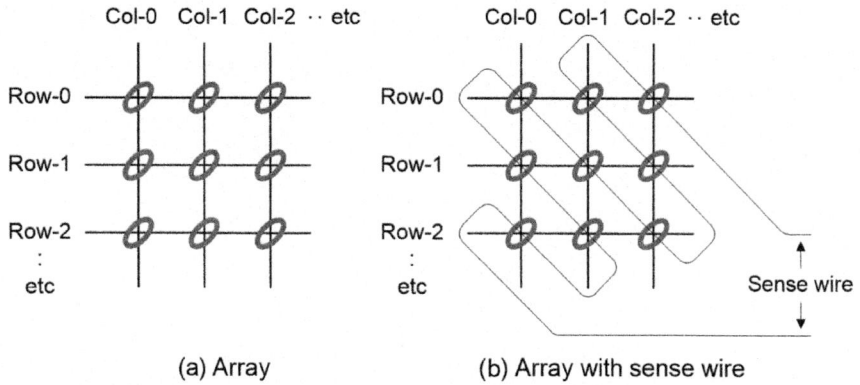

Ferromagnetic beads organized as an array

Fig. 8.16 Magnetic core store (Courtesy: Clive Maxfield). *Source* https://www.clivemaxfield.com/diycalculator/popup-m-hrrgcomp.shtml

Writing is accomplished by passing just enough current create ½ the critical magnetic field through an X and Y wire. This will cause the field strength at the crossing point to produce a flux density greater than the cores saturation, thus magnetizing this particular core. Ones and zeros are represented by the direction of the field, which can be set simply by changing the direction of the current flow in one of the two wires.

A third wire-the sense/inhibit line-is needed to read a bit. Reading uses the process of writing; the X and Y lines are powered in the same fashion that they would be to write a "0" to the selected core. If that core holds a "1" at that time, a short pulse of electricity is induced into the sense/inhibit line. If no pulse is seen, then the core held a "0". This process is destructive; if the core did hold a "1", that pattern is destroyed during the read, and must be re-set in a subsequent operation.

Magnetic core storage was largely replaced by other non-volatile systems in the form of magnetic media (8″, 5½″, 3¼″ floppies, tapes, drums and hard discs) with most other memory being still volatile (SRAM, DRAM etc.).

Since the advent of the microprocessor, the tendency has been toward semiconductor memories. The most commonly used are:

ROM: Read only memory–like hard wired matrices.
RAM (static): Flip-flops.
RAM (dynamic): Charge storage elements which must be refreshed.

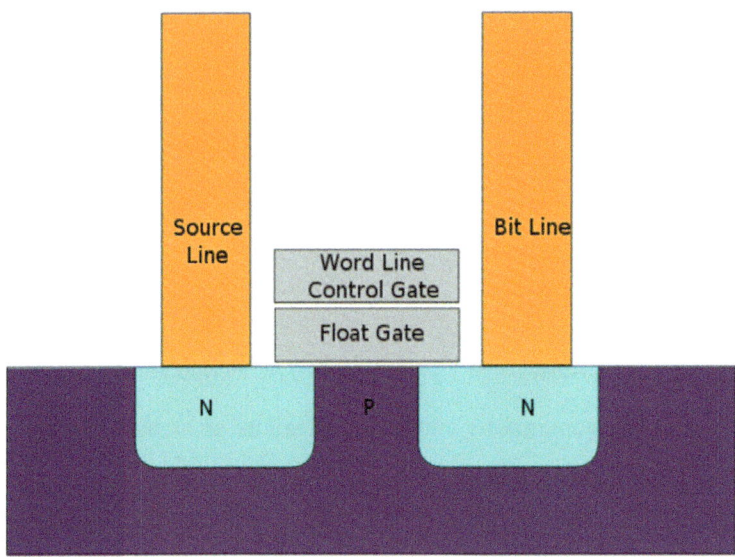

Fig. 8.17 Flash Memory. *Source* https://en.wikipedia.org/wiki/Flash_memory

PROM: Programmable read only memory-similar in function to punched cards.
EPROM: Electrically programmable read only memory–reprogrammable PROM.
EEPROM: Electrically erasable programmable read only memory.
SDRAM: Synchronous Dynamic Random Access Memory.

SDRAM are currently the best non-volatile alternatives to magnetic media and form the basis of flash memory. Today much long term non-volatile storage is carried out with flash memory (USB stick). NAND flash uses floating-gate transistors which are connected in a way that resembles a NAND gate (Fig. 8.17).

A floating gate transistor resembles a standard MOSFET, except that the transistor has two gates instead of one. On top is the control gate (CG) as in other MOS transistors, but below this there is a floating gate (FG) completely insulated by a surrounding oxide layer. The FG is interposed between the CG and the MOSFET channel. Because the FG is electrically isolated by its insulating layer, electrons placed on it are trapped until they are removed by another application of an electric field.

8.6 A Compact History of the Microprocessor

Ostensibly, the first single chip microprocessor, the 4004 was released by the Intel Corporation in November 1971. Using MOS LSI technology, the chip marked both a technological and economic milestone in computing [2]. The remaining historical development can be seen from the list below.

Date	Firm	Microrpocessor	Data	Clock	Type
1970	Garrett	AiResearch MP944	20 Bit	375 kHz	Secret until 2013*
1971	Intel	4004	4 Bit	108 kHz	
1972	Intel	4040	4 Bit	740 kHz	
1972	Intel	8008	8 Bit	299 kHz	
1974	Intel	8080	8 Bit	2 MHz	
1974	Texas	TMS1000	4 Bit	400 kHz	microcontroller
1974	AMD	2900	8 Bit	1 MHz	bit slice
1974	Fairchild	F8	8 Bit	2 MHz	
1975	Motorola	6800	8 Bit	2 MHz	
1976	Zilog	Z80	8 Bit	2.5 MHz	
1976	Nat Semi	SC/MP	8 Bit	2 MHz	
1976	Ferranti	F100-L	8 Bit	8 MHz	military
1976	Intersil	6100	12 Bit	3 MHz	PDP8 clone
1976	Texas	TMS9900	16 Bit	4 MHz	
1976	Zilog	Z8000	32 Bit	10 MHz	
1977	Motorola	6502	8 Bit	2.2 MHz	
1977	AMD	2901	4 Bit	6 MHz	bit slice
1977	Intel	8051	8 Bit	8 MHz	microcontroller
1978	Signetics	2650	8 Bit	2 MHz	
1978	Intel	8086, 8088	16 Bit	5 MHz	pipelined
1979	Motorola	6809	16 Bit	2 MHz	
1979	Motorola	68000	32 Bit	8 MHz	
1980	Intel	8087	16 Bit	10 MHz	math coprocessor
1982	Intel	80186/80286	16 Bit	6 MHz	
1982	Intel	80187	16 Bit	6 MHz	math coprocessor
1982	Intel	80286	16 Bit	8 MHz	
1982	Intel	80287	16 Bit	8 MHz	math coprocessor
1983	Hitachi	6301	8 Bit	8 MHz	
1983	Motorola	MC14500	1 Bit	1 MHz	bit slice
1983	INMOS	T212	16 Bit	5 MHz	parallel
1984	Motorola	68881/68882	32 Bit	12 MHz	math coprocessor
1984	INMOS	T414	32 Bit	12 MHz	parallel
1985	INMOS	T800	64 Bit	12 MHz	parallel
1985	Intel	80386	32 Bit	16 MHz	
1986	Motorola	68030	32 Bit	20 MHz	
1987	Intel	80387	32 Bit	16 MHz	math coprocessor
1988	Motorola	88000	32 Bit	20 MHz	
1988	Texas	TMS320C30	32 Bit	27 MHz	
1989	Intel	80486	32 Bit	25 MHz	
1989	Intel	80487	32 Bit	25 MHz	math coprocessor
1992	Intel	Pentium	32 Bit	66 MHz	
1995	Intel	Pentium Pro	32 Bit	200 MHz	
1995	INMOS	T-9000	64 Bit	50 MHz	parallel
1996	Intel	80587	32 Bit	60 MHz	math coprocessor

1997	Intel	Pentium II	32 Bit	233 MHz
1998	Intel	Pentium II Xeon	32 Bit	450 MHz
1999	Intel	Celeron, Pentium III	32 Bit	450 MHz
2000	Intel	Pentium IV	32 Bit	1.5 GHz
2001	Intel	Xeon, Itanium	64 Bit	1 GHz
2002	Intel	Celeron	64 Bit	1.3 GHz
2004	Intel	Celeron M	64 Bit	2.0 GHz
2005	Intel	Pentium D	64 Bit	3.2 GHz
2006	Intel	Dual Core Xeon-7140M	64 Bit	3.33 GHz
2007	Intel	Pentium Dual-Core	64 Bit	2.7 GHz
2008	Intel	Quad Core Q9300	64 Bit	2.5 GHz
2009	AMD	Quad Core Phenom II X4	64 Bit	3.7 GHz
2010	AMD	Hex Core Phenom II X6	64 Bit	3.7 GHz
2012	AMD	A10-5800K	64 Bit	4.2 GHz
2013	Oracle	16 Core SPARC T5	64 Bit	3.6 GHz
2015	IBM	8 Core z13	64 Bit	5.0 GHz
2017	IBM	10 Core z14	64 Bit	5.2 GHz

*Ostensibly invented by Steve Geller and Ray Holt for the US Navy F14 "Tomcat" fighter aircraft. <http://www.firstmicroprocessor.com/thepaper/>

Though not exactly a single chip system, the microprocessor developed by Geller & Holt predated Intel's 4004. Secret until 2013, it serves as a good example of military developments which first become known to the public much later

References

1. Chase. G.C.—History of Mechanical Computing Machinery—Annals of the History of Computing Vol. 2, No. 3 July 1980.
2. Faggin. F., M.E. Hofflr., S. Mazor & M. Shima – The History of the 4004 – IEEE Micro, December 1996.
3. Garratt. G.R.M.—The Early History of Radio: from Faraday to Marconi—Institution of Engineering and Technology History of Technology, 1994.
4. Gregersen, Erik. "Ada Lovelace: The First Computer Programmer". *Encyclopedia Britannica*, Invalid Date, https://www.britannica.com/story/ada-lovelace-the-first-computer-programmer. Accessed 2 March 2024.
5. Ifrah, Georges. (2001). The Universal History of Computing: From the Abacus to the Quantum Computer. John Wiley and Sons, Inc: New York. pp121-133.
6. Kwa Tak Ming – The Fundamental Operations in Bead Arithmetic—How to Use the Chinese Abacus—France Press, 20. April 2011. {ISBN-10: 1447401956}
7. https://archive.computerhistory.org/resources/access/text/2016/12/B1671.01-05-01-acc.pdf
8. Pechenkin, A. (2019). The Mandelstam School: Theory of Non-linear Oscillations. In: L.I. Mandelstam and His School in Physics. Springer, Cham. https://doi.org/10.1007/978-3-030-17685-3_9
9. Zhou. F, J. Chen, X. Tao, X. Wang and Y. Chai—2D Materials Based Optoelectronic Memory: Convergence of Electronic Memory and Optical Sensor (Review Article) – AAAS Research, 2019 {https://doi.org/10.34133/2019/9490413}

Internet Sites

10. https://chamer-rundfunkmuseum.de/

11. https://www.livescience.com/20718-computer-history.html
12. https://www.computerhistory.org/
13. http://www.lindnilsson.dk/lars/en/calculators/2/Comptometer%20-%20Model%20H
14. http://www.computerhope.com/issues/ch000984.htm
15. https://www.calculemus.org/MathUniversalis/1/MU1_1-3.html
16. http://www.rechenmaschinen-illustrated.com/Tchebyshev.htm
17. https://www.didaktik.mathematik.uni-wuerzburg.de/history/ausstell/leibniz/rechenmaschine.html
18. http://www.howstuffworks.com/microprocessor1.htm
19. http://parallel.ru/history/wilson_history.html
20. https://en.wikipedia.org/wiki/Microprocessor_chronology